# Conflicting Structures

## Vladimir A. Lefebvre

Leaf & Oaks Publishers
Los Angeles

First publication: Vysshaya Shkola, 1967 (in Russian)
Second edition: Sovetskoe Radio, 1973 (in Russian)

Translation by Victorina D. Lefebvre

The first edition of this book appeared in 1967 (in Russian). In that edition, the author introduced two completely new concepts: that of a reflexive system (a system that has an image of the self) and that of reflexive control (conveying a basis for making the decision that is advantageous to the side conducting the reflexive control); both concepts have since become firmly established in modern theories of decision-making. The book contains the author's model of the Universe as a reflexive system (Janus-Cosmology) as well as the description of a device that turns fears into reality through reflexive control, constructed by the author for the purpose of experimental study. In addition, the author also explains how to use reflexive control over processes of reflexive control.

This translation into English has been made from the second revised edition of 1973, which includes more detailed descriptions of the experiments conducted.

*Key words*: reflexive system, reflexive control, reflexive control over reflexive control, self-reflexion, coordination, focal points, feedback, military conflict, cosmology, relation tissue-pattern

ISBN  978-0-578-15769-6

# CONTENT

# Introduction

The tradition of the natural sciences as it has evolved through the twentieth century contains two hidden postulates at its core. The first one says: *the theory of an object is not created by the object itself.* The second says: *an object does not depend on the existence of the theory of the object.* The tradition of natural sciences has appeared in a battle with theological concepts. The first postulate emphasizes the researcher's dominant position in relation to the object: there are no objects superior to the researcher; objects cannot penetrate his designs, nor can they either hinder or help him to investigate them. The thesis that nature is not malevolent is a consequence of this postulate. In its essence, this postulate denies the existence of the afflatus. The second postulate allows us to speak of properties and laws as inherent in objects. Such properties exist objectively and are only articulated by the researcher.

If we accept the possibility that a scientific theory may influence its object, then it follows that a theory of the object's properties may change those properties and even undermine the veracity of the theory itself. The situation we are considering here differs in principle from the familiar problem of quantum physics: that the knowledge of an object in that field depends on and, indeed, is produced by the object's interaction with an instrument of study. Knowledge itself, namely by the fact of its articulation, has no effect on the nature and properties of the physical processes described in the theory; for example, the value of the Planck constant does not depend on the fact that it has been made public. Note that, since a scientific theory is not a physical object, it cannot be considered to be part of an instrument of study.

The two postulates formulated above belong to the framework of the physical sciences. The situation with respect to the research of mental phenomena is different: it has a long history in which theological, psychiatric, and "hard" scientific lines of inquiry are interwoven.

At the end of the nineteenth and the beginning of the

twentieth century, attempts were made to turn psychology into a "hard science," following the example of the disciplines that had achieved the most success, especially physics.

The study of mental phenomena is fundamentally different from that of physical phenomena. The "given" of the physicist is a huge, resplendent world of things, whereas the "given" for the psychologist is a dark inner world, intangible and ephemeral. When a psychologist looks inside the "self," he tries to represent this "given" by means of carefully developed concepts whose synthesis (successful or unsuccessful) serves to represent "mental experience."

The act of self-analysis with the purpose of obtaining information about one's own mental world has been termed introspection. The next step in the psychologist's work consists of projecting the results of his self-observation onto the thoughts and feelings of other individuals.

The mental world of others is concealed from the investigator. He can observe only behavior. Yet since his own behavior is generated by his own mental world, he supplements the other's behavior with constructions of the mental worlds of his subjects. This is the proper work of the psychologist. Different schools of psychology may interpret it differently, but the essence of the task remains the same.

At the beginning of the twentieth century, the methods of psychological research changed radically; the researchers deliberately abandoned introspection. The American psychologist E.L. Thorndyke, proposed a rejection of introspective analysis in favor of "actual givens," that is, behavior. This was the foundation of behaviorism, an approach that promised to make psychology truly "scientific." The behavioristic approach developed the concept of *a black box* having input (stimulus) and output (response). Stimuli can be registered and measured, and responses can be registered and measured. The psychologists' new task was to find functional correspondences between input and output. In this way, anything resembling "mental phenomenology" remained outside the field of scientific analysis.

The behavioristic scheme contains one striking contradiction. Suppose that an investigator uses a text as a stimulus. The text has the meaning, and it is obvious that a subject's response is to a large degree determined by the meaning of the text. Yet the text obtains its meaning only in the mental worlds of the subject and the experimenter. As a physical object, the text is merely a sequence of symbols or sounds. Thus, to define the stimulus, the experimenter must first consults his own mental world, that is, commit an act of introspection. Moreover, the experimenter must assume that the subject understands the meaning, that is, he must commit an act of projection granting the subject a separate but correlative mental world.

The stimulus-response scheme that has castrated psychology does not permit the researcher so much as to register the fact that the subject was deceiving during an experiment; even this registration requires the experimenter to assume sympathetic knowledge of the subject's mental world. Thus, rejecting introspection leads either to the absence of substantive research or to using introspection as a tacit, unacknowledged methodological component.

Besides behaviorism, there have been other attempts to find patterns of mental activity while avoiding the use of introspection. None, however, has allowed us to avoid introspection entirely. The core of the matter is that physiological states must be interpreted. For example, it is impossible to register pain with a measuring device; the device measures only some physical parameter, which is then interpreted by the experimenter as "pain." The researcher is forced to consult his own, individual mental experience. It is clear that here we are using a "backroom measuring technique." Using physiology, however, does not eliminate the necessity of studying mental phenomena as such. We are brought back to the classic philosophical problem of the relation between mind and its material substratum.

The particular ways of relating the concepts of mind and matter have always been determined by a broader intellectual culture. There have been models based on the relation part-whole, others on the relation of permeation, and others that, from today's point of view, seem quite primitive. For example, one might be subordinated to the other: "Although the soul has no form it acquires (as light does) the extent and the form of the body in which it lives" (Chatterjee and Datta, 1954).

Today we have a sophisticated intellectual culture. Many traditional philosophical problems have obtained a new lease on life by creating their own disciplines. For example, the classical problem of the Universe - is it finite or infinite? - is now studied within the framework of the general theory of relativity. Discussing this problem in the terms of ordinary reasoning is simply of no interest.

The problem of the relation between mind and matter has a different status. For a modern engineer, the mere existence of mental phenomena is not entirely obvious. He recognizes only those phenomena which, at least potentially, can be understood in behavioristic terms. The triumph of the cybernetic approach has not only meant the appearance of new and efficient means for the analysis of complex systems, but also brought about a cataclysmic reduction of the ontological field within which scientific analysis takes place.

Ironically, the problem "can artificially created things be endowed with a mental domain?" has turned into the problem "can computers think?" Turing's solution to the latter problem was in the best traditions of behaviorism.

It seems to us that a principal methodological issue for the study of complex objects is the elaboration of models of reality, models in which mental and material phenomena may be structurally connected. Depending on how this issue is resolved, we would either consider systems endowed with intellect to be part and parcel of physical systems or, conversely, we would be satisfied with two parallel lines of study and formulate our failure in a way similar to Bohr's principle of complementarity (Bohr, 1955).

In this book, we offer some methods that may be useful in solving problems requiring the synthesis of different approaches.

In any case, mental phenomenology is unescapable. By trying to avoid or deny it, the researcher inevitably reduces the complexity of the complex systems under analysis. By accepting the "real" existence of mental phenomena, on the other hand, the researcher accepts the existence of an object, comparable to or even excelling himself in perfection; in objectifying mental experience, the researcher must then admit the varying degrees of its intensity. In this way, the researcher rejects the two postulates of methodological tradition of the natural sciences.

Drawing an absolute distinction between objects and the researcher is justifiable only for objects not endowed with a mental world. In the case of objects endowed with mental experience, the relation between the researcher and the object turns into a relation between two researchers, each of which is an object of inquiry for the other (Lefebvre, 1969).

In particular, the relation between object and researcher is likely to manifest itself in terms of a conflict. This conflict is of paramount interest for the analysis of relationships between the researcher and systems that may equal or excel him in perfection. The penetration of an adversary's plan, i.e., knowledge of his thoughts, becomes critical. The objective situation forces the participant in a conflict to study the adversary's mental world and construct a "theory" of it. This is a very special but important case of object-theory interaction. The object seeks to disprove the theory constructed for him; he constantly deviates from the theory and renders the theory continuously incorrect.

Thus, in a situation of conflict, the second natural-scientific postulate is violated. The first postulate is also violated when one of the adversaries imposes on the other a certain image of the self.

By investigating social-psychological phenomena, the researcher becomes one participant in a game, what we will call a reflexive game. Since the researcher cannot rule out the possibility of contact with the participants that are under investigation, his theoretical constructs, if understood by the other participants, may drastically change the functioning of the entire system. The researcher may even find himself trapped by the theory another participant constructs of him.

An interesting facet of this general case is reflexion. In its traditional philosophical or psychological meaning, reflexion is the ability to take the attitude of an observer, an investigator, an examiner with respect to one's own body, one's own actions and thoughts. We will broaden this concept and consider reflexion as including the ability to assume the position of an investigator of another person's actions and thoughts.

This broadened concept of reflexion allows us to construct a holistic subject of investigation and consider reflexive processes as a single phenomenon with the peculiarity that the researchers play the role of objects to each other.

Note that penetration into the mental world of the other may be accomplished by a psychologist, for whom this action is an end in itself, or by any person engaging in natural communication with another person. The structure of the two is the same; they differ only in their purpose and goals.

The author's task in this work is to make reflexive processes the object of specialized analysis.

## Acknowledgment to the English Edition

I am grateful to Victorina Lefebvre for translating this book from Russian into English. In addition, she took upon herself the technical preparation of the manuscript for print. I am also thankful to Harold Baker for his linguistic advice and corrections.

# Chapter I.

## The Algebra of Reflexive Processes

What is a reflexive system? Let us use the following analogy. Imagine a room full of crooked mirrors placed at various angles to each other, as is typical in amusement parks. If a pencil falls from a table, it will be reflected by the mirrors in many whimsical ways; then the reflections will be reflected with numerous distortions, *ad infinitum*. An avalanche of distorted images will flash around the room. A reflexive system is a system of mirrors reflecting each other over and over again. Each mirror is analogous to a person with a particular position relative to the world. The entire complicated stream of the mirrors' mutual reflections is an analogue of the reflexive process.

The example with the pencil illustrates the difference between physical processes and social-psychological ones. A pencil falling is a physical process. If, however, we are interested not only in this fall but in the entire stream of multiple mirror images, we are dealing with a social-psychological event.

Imagine that an observer enters this room. (An observer may be considered a special kind of mirror.) The entire situation changes fundamentally. Each movement by the observer-mirror will be accompanied by corresponding changes in the multiple reflections.

Sometimes we will speak about an *external* observer, implying that the mirror-persons whom he investigates do not reflect him.

This is the starting point of our construction. One should not take the analogy too literally; it is only an illustration. Further on we will introduce a special apparatus designed for studying reflexive processes, and we will use human conflict as an example. This does not mean that the apparatus is only for the analysis of conflict, merely that reflexive processes are seen more clearly in such situations.

## Representation of Reflexive Systems

Symbols $X, Y, Z$ represent the conflicting adversaries. In order to make a decision, $X$ constructs a model of the situation: the arena where the interactions take place, together with the participants and their troops. $Y$ also constructs a model of the situation; in addition, $Y$ may realize that $X$ has his own model of the situation. In turn, $Z$ realizes that $X$'s and $Y$'s mental worlds include models of the situation. Success in a conflict is largely determined by the way the adversaries represent each other's mental world. Without a detailed model, in which the peculiarities of the reflexive structure of an adversary's mental world are taken into consideration, it is impossible to interpret the adversary's actions correctly. For example, a movement on the ground may have a real purpose or may be just intended to influence the adversaries' decisions. Even with a small number of participants reflexive processes have a complicated structure and special methods are necessary to make them an object of analysis.

Let us depict an arena where three people are interacting as a rectangle containing three circles (Fig. 1).

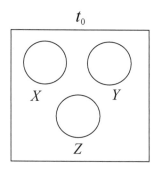

$t_0$

Fig. 1

At moment $t_1$, $X$ becomes aware of the situation; this means that he acquires a mental image of the arena. The picture in Fig.1 is moved inside $X$ (Fig. 2). It is obvious that the entire system has changed; it has acquired new elements.

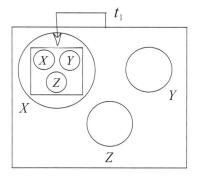

Fig. 2

Then at moment $t_2$, Y becomes aware of the new situation. To depict this process, we transfer the image in Fig. 2 inside circle Y (Fig. 3).

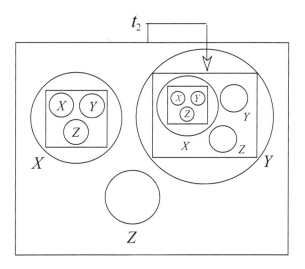

Fig 3

If, at moment $t_3$, Z becomes aware of the situation, we would have to move the entire picture given in Fig. 3 inside circle Z. It is hard to do, for purely graphical reasons. Also, it would be inconvenient to operate with such depictions. Rather we introduce an algebraic language that allows representation of reflexive processes of arbitrary complexity.

Symbol $T$ represents the arena where $X$, $Y$, and $Z$ interact. Figure 1 corresponds to this symbol. The participants' images of the arena are represented as $Tx$ ($T$ from $X$'s point of view), $Ty$ ($T$ from $Y$'s point of view), and $Tz$ ($T$ from $Z$'s point of view). Elements $Tx$, $Ty$, and $Tz$ appear as the results of processes of awareness. Figure 2 depicts the case of $X$'s awareness. Similar pictures can be drawn for the other two persons. An image belonging to one person can be reflected by the other one so that the following elements appear: $Txy$ ($Tx$ from $Y$'s point of view), $Txz$ ($Tx$ from $Z$'s point of view), $Tyz$ ($Ty$ from $Z$'s point of view) etc. The elements with two indexes can also be reflected: $Txyz$ ($Txy$ from $Z$'s point of view), $Txzy$, $Tzxy$, etc. The image a person had at moment $t_1$ can be reflected by him at moment $t_2$ as an image rather than as physical reality. As a result we obtain the elements $Txx$, $Tyy$, $Txxx$, etc.

Now we will represent the process of the three persons' interactions in the arena. At moment $t_0$ the persons do not yet have images (Fig. 1). This system is depicted as $\Omega_0=T$. The reflexive system in Figure 2 is represented as the sum[1] of two components: the arena and $X$'s image of the arena:

$$\Omega_1 = T + Tx . \tag{1}$$

The system in Figure 3 corresponds to the polynomial

$$\Omega_2 = T + Tx + (T + Tx)y , \tag{2}$$

where the expression in parentheses is $Y$'s image of the picture in Figure 2 moved inside circle $Y$ (Fig. 3). This notation eliminates the difficulties attending graphical representation of such systems and their description in natural language. The reflexive system after $Z$ becomes aware of its previous state, is represented as

$$\Omega_3 = T + Tx + (T + Tx)y + [T + Tx + (T + Tx)y]z . \tag{3}$$

It seems natural to apply the distributive law to the right-hand index, which will allow us to remove the parentheses. For

---

[1]Symbol '+' will be justified by its formal terms.

example, the following two expressions are equivalent:

$$T + Tx + (T + Tx)y = T + Tx + Ty + Txy .$$

This law can be interpreted in two ways. From the external observer's point of view, the element $(T+Tx)y$ - the result of taking index $y$ out of the brackets - represents Y's mental world as a whole. This does not mean that Y, himself, has this very picture. On the other hand, this element may mean that Y does see the world in such way, i.e., that the operation of taking the index out of the brackets is going on in Y's mental domain.

We will also allow reproduction of a summand without violating the equivalency of polynomials, because a person (or an observer) does not obtain new information as a result of replicating the text already known. For example:

$$T + Tx = T + Tx + Tx .$$

Let us note that the expression above does not say anything about the adequacy of persons' reflections of other persons' mental images. For example, consider two elements, $Tx$ and $Txy$. Person Y may have either a correct image of $Tx$ or an incorrect image. The notation registers only the fact of this element's existence in Y's mental world. Additional commentary is necessary to determine its degree of adequacy from the external observer's point of view.

## Operators of Awareness

Let us introduce now special notation to register the process of awareness. To do so, we have to find a formal way to represent the transition from $\Omega_0$ to $\Omega_1$, $\Omega_1$ to $\Omega_2$, from $\Omega_2$ to $\Omega_3$, etc. Polynomials (1), (2), (3) differ fundamentally from common polynomials with real coefficients, and it is necessary to define the algebraic object with which we will be working. For three people, the initial symbols are $x$, $y$, $z$, $T$, and 1. A finite sequence of these symbols constitutes a word, for example, $x$, $xy$, $Tx$, $xyz$, etc. Two words are considered equivalent if they differ only by the presence of symbol 1 (for example, $x1xy1=xxy$). Thus, we may eliminate symbol 1 from words without changing their meaning.

Consider words that do not contain the symbol $T$. They constitute a denumerable set, and after numbering them we obtain a sequence $a_i$. Now we can define our concept of polynomial. The following symbolic sum

$$\omega = \sum_{i}^{\infty} \alpha_i a_i ,$$

where $\alpha_i$ is an element of Boolean algebra with values 0 and 1, will be called *polynomial.*

With a given numeration of $a_i$, a polynomial is unambiguously set by coefficients $\alpha_i$. From now on, we will write only those elements that have the coefficient 1. Note the difference between a polynomial and a single word. For example, if we write $\omega=1$, this means that we consider polynomial $1 + \sum_{i=2}^{\infty} 0 a_i$ , where only element $a_1=1$ has a coefficient not equal to zero.

The addition and multiplication of polynomials are similar to those for common polynomials, except that multiplication is not commutative; the associative law and left and right distributive laws hold:

$$\omega_1(\omega_2 + \omega_3) = \omega_1\omega_2 + \omega_1\omega_3$$
$$(\omega_2 + \omega_3)\omega_1 = \omega_2\omega_1 + \omega_3\omega_1.$$

We put each polynomial $\omega$ in correspondence with a specific polynomial $\Omega=T\omega$. As was shown earlier, polynomials $\Omega$ represent states of reflexive systems, and polynomials $\omega$ will be interpreted as *operators of awareness.*

Now we will show the procedure for converting the picture in Figure 1 into the picture in Figure 2, etc., in algebraic language. To do so, we right multiply polynomial $T$ (expressing the content of Figure 1) by polynomial $1+x$:

$$\Omega_1 = T(1 + x) = T + Tx. \tag{1'}$$

To move to the state $\Omega_2$, we need to right multiply polynomial $\Omega_1$ by $(1+y)$:

$$\Omega_2 = \Omega_1(1 + y) = T(1 + x)(1 + y) = T + Tx + (T + Tx)y. \tag{2'}$$

State $\Omega_3$ is generated by right multiplication of $\Omega_2$ by $(1+z)$:

$$\Omega_3 = \Omega_2 (1 + z) = T(1 + x)(1 + y)(1 + z)$$

$$= T + Tx + (T + Tx)y + [T + Tx + (T + Tx)y]z. \qquad (3')$$

Thus, the process of awareness that was depicted graphically as the schematization of a natural intuitive understanding of reflexion now corresponds to the algebraic operation of multiplication of a polynomial by $(1+x)$, $(1+y)$, $(1+z)$. This is a case of the participants' consecutive awareness. It is easy to show how all three persons act simultaneously. The operator of awareness is $\omega=1+x+y+z$, and evolution of the reflexive system is expressed by $\Omega n=T(1+x+y+z)^n$, where $n$ is the number of acts of awareness.

Such representation of processes of awareness significantly broadens our ability to study complicated types of awareness, including those practically inexpressible in graphical or natural language.

## Operator Generating the MaxMin Principle

The MaxMin principle developed by von Neumann underlies the contemporary ideology of decision making. It means that a decision maker must guarantee a minimal loss for the self. Let us look at the reflexive structure of players who exemplify this ideology. $X$ has to make the best possible decision, such that with any other $X$'s decision, an opponent may increase $X$'s losses. Suppose that $X$ has no theory that would allow him to make decisions without thought. He examines each possible decision and compares it with the best responding decision by his opponent. In this way, the opponent is present in $X$'s mental world and constantly watches over his thoughts. This player can be represented by the following polynomial:

$$\Omega^* = T + (\Omega + \Omega y)x . \qquad (4)$$

$X$'s mental world is such that any image, including his image of the self, is (from his point of view) adequately reflected by his opponent. For that reason, any thought that $X$ is aware of, is reflected by $X$'s opponent[2]. If $X$ is in conflict with $Y$, this structure of his mental world leads to using the MaxMin principle, i.e., $X$ must make decisions such that the adversary, while knowing them and making the best decisions of his own, would minimally damage $X$. In many conflicts, however, there is no such optimal decision: all decisions are unsatisfactory. This forces a player to neutralize the adversary's deduction; he must make decisions without reasoning, i.e., cast lots. By reading $X$'s mind, the adversary will not be able to deduce $X$'s decision (the result of throwing dice is impossible to predict), but will realize that $X$ is using randomization.

A player who uses the MaxMin principle is represented by expression (4). We assume that a polynomial may change only as a result of the act of awareness. If we apply operator $\omega = 1 + x$ to (4), as we did in examples (1'), (2'), (3'), polynomial (4) will change its structure. But we want player $X$ to hold on to the MaxMin principle even while performing acts of awareness, that is, the type of the polynomial must be invariant to the act of awareness.

The simplest operator of awareness that generates and preserves a mental world similar to the one expressed by (4) is as follows:

$$\omega = 1 + x + yx. \qquad (5)$$

After using it we obtain:

$$\Omega^*(1 + x + yx) = [T + (\Omega + \Omega y)x](1 + x + yx)$$

$$= T + \Omega x + \Omega yx + \Omega^* x + \Omega^* yx = T + [(\Omega + \Omega^*) + (\Omega + \Omega^*)y]x$$

$$= T + (\Omega' + \Omega'y)x,$$

i.e., $X$'s mental world may change as a result of acts of awareness but, from $X$'s point of view, $Y$ continues to play the dominant role and to know all of $X$'s thoughts:

$$\Omega^*\omega = T + (\Omega' + \Omega'y)x.$$

Thus, operator (5) keeps the structure of polynomial $\Omega^* = T + (\Omega + \Omega y)x$

---

[2] Of course, this assumes that the image $\Omega$ that $X$ has is identical to the image $\Omega$ that, from $X$'s point of view, $Y$ has.

unchanged.

If the only operator of awareness that $X$ has is $\omega = 1 + x + yx$, then this person is represented by polynomials of type (4) and is doomed to persevere with the MaxMin principle. Such a person is closed by this operator. Its multiple use does not change the structure of the polynomial.

Operator (5) generates a special reflexive closure. $X$'s awareness that he is designed in this particular way changes his mental self-image, but nevertheless $Y$ remains a peculiar all-seeing eye immediately reflecting $X$'s new image of the self. This awareness does not eliminate the all-seeing eye, which keeps its dominant position. $X$ may adequately reflect his own structure, but this fact, from $X$'s point of view, will simultaneously be reflected by $Y$.

Note that the polynomial may evolve through serial acts of awareness without any information from the outside. All new information appears as a result of reflecting the previous state. In other words, the operator generating the MaxMin principle is a special form of self-awareness.

We may suppose that this operator underlies some types of religious thinking. The God of the Calvinists is an "all-seeing eye," aware of all the individual's thoughts. Since the work of the operator of awareness is not governed by a person, this may lead to paradoxical and painful states of awareness in the believer. Even when the person considers himself a nonbeliever, this self-image is "seen" by the "all-seeing eye," that is, God is still present in the person's mental world.

To clarify the work of this operation of awareness, we construct a model of $X$ with his "mental display" (the square in Fig.4). $Y$, although being outside of the square, is perceived by $X$. The content of the display is perceived by $X$ via two channels: on the one hand, directly from the display, and on the other, indirectly through the $Y$-image, which cannot be removed by any act of awareness. If the situation depicted in Fig. 4 is reflected in the mental display, this will not change the structure of the process of awareness (Fig. 5), just as a picture of a movie projector on the screen does not change the work of the projector.

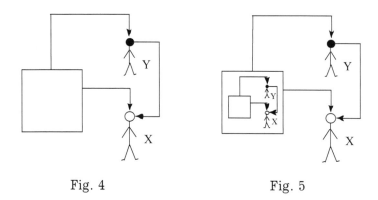

Fig. 4                              Fig. 5

This graphical representation of the process of awareness, although it does not show certain subtle features of the process, roughly depicts events not captured by the algebraic apparatus. For example, we may "draw" a "special design" on the display, which, from the person's position, is indistinguishable from the projected image. From the external observer's position, only part of the content is the result of reflecting, while the participant does not distinguish between the elements drawn on the display and those reflected.

## Reflexive Closures of Other Types

The following identity represents the invariance of a polynomial to the operator of awareness:

$$(T + \Omega\omega)(1 + \omega) \equiv T + \Omega'\omega ,$$

where $\Omega' = T + \Omega + \Omega\omega$.

Consider the operator

$$\omega = 1 + x^2.$$

One application to $T$ generates the polynomial

$$\Omega_1 = T + Txx.$$

$X$ has no image of the real arena but only of the arena from his own point of view. This is a case of a solipsistic mental world. For $X$, reality is only an element of his mental world. The act of awareness of his own state with the operator $\omega=1+x^2$ leads again to the solipsistic mental world, i.e., this type of mental world is "locked-in" with respect to the given operator:

$$\Omega_1 = (T + Txx)(1 + x^2) = T + \Omega'xx \ .$$

The operator of awareness $1+x^2$ dooms a person to relate to reality as an element of his own mental world. If such person is an external observer, then $T$ will be absent in the polynomial representing him (see Fig. 6). There is no direct channel from the mental display to $X$, only a channel from $X$ to $X$.

Consider the operator

$$\omega = 1 + yx.$$

One application generates the polynomial

$$\Omega_1 = T + Tyx.$$

The image of $X$'s reality is a phenomenon inside $Y$'s mental world (see Fig. 7). This pathological state is also closed, because the following expression holds:

$$(T + \Omega yx)(1 + yx) = T + \Omega'yx.$$

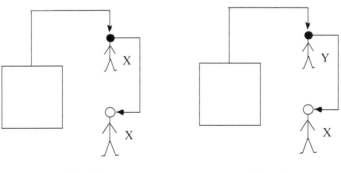

Fig. 6          Fig. 7

For $X$, the direct channel of realizing the content of the display is absent: as in the solipsistic case, the channel goes through $Y$.

Consider the operator

$$\omega = 1 + x + x^2.$$

A person with this operator performs double awareness: the fact of awareness is also reflected (Fig. 8).

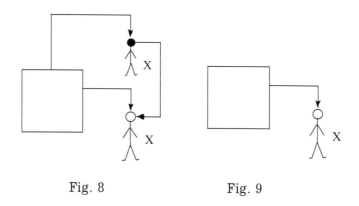

Fig. 8                                   Fig. 9

Figure 9 corresponds to the simplest operator:

$$\omega = 1 + x.$$

Let us consider a more complicated operator of awareness, which will be used later.

$$\omega = 1 + x + yx + zx + yzx.$$

Its multiple applications generate polynomials of the type

$$\Omega' = T + [\Omega + \Omega y + (\Omega + \Omega y)z]x.$$

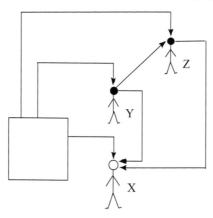

Fig. 10

From $X$'s point of view, any image or thought that he is aware of is reflected by $Y$; person $Z$ also reflects any thought and image that $X$ is aware of and, in addition, reflects the fact of $Y$'s reflecting $X$'s thoughts and images (Fig. 10). $Y$ and $Z$ may be in various positions of subordination, but these positions cannot be removed by an act of awareness.

Perhaps this kind of operator of awareness is at work in the Catholic and Orthodox mind. God is $Z$, and a priest is $Y$. Confession is the means for preserving this operator of awareness. From $X$'s position, $Y$ being actually present dominates $X$'s mental world. In the preparation for confession and during it a person's mental world is verbalized and becomes available for domination. In this situation, $Y$'s function is to activate $X$'s process of self-awareness, because without images in $X$'s mental world their reflection by $Z$ is impossible.

## Development of a Polynomial

An algebraic approach to reflexive structures generates particular problems. For example, can a system move from state $\Omega_1$ to state $\Omega_2$? This question can be reduced to the question of whether equation $\Omega_1\omega=\Omega_2$ has a solution. This equation, nonlinear in relation to $\omega$, may have multiple solutions or no solution. For example, equation $(1+x)\omega =1+ x + x^2 + x^3$ has two solutions: $\omega_1=1+ x + x^2$ and $\omega_2=1+x^3$, while equation $(1+x)\omega=1+x^3$ has no solution.

Prior to this point we have assumed that a person has only one operator of awareness. Now we will drop this assumption and allow a person to have a set of operators. In the framework of our specific construction, we may pose the question of the history of some state $\Omega$. To do so, $\Omega$ must be represented as a product of co-factors:

$$\Omega = T\omega_1\omega_2 \ldots \omega_k.$$

Since such factorization may be not unique, we may receive more than one trajectory of the sequential operators' work. There is special interest aroused by factorization of irreducible factors-polynomials that cannot be represented as a product of two polynomials each not equal to 1. Irreducible factors are interpreted as elementary acts of awareness. Note that in the calculus being constructed here a theorem of the unique factorization of irreducible factors does not hold. For example, there are two ways to represent the polynomial $(1 + x)\omega = 1 + x + x^2 + x^3$ :

$$\omega = (1 + x)^3 \text{ and } \omega = (1 + x)(1 + x^{\,2}).$$

Of course, such reconstruction of history has a meaning only in the framework of this model, with all its limitations, the most important of which is making a factor the analogue to awareness. The factorization described above is only a particular case and the simplest one. Further we will demonstrate other ways of developing polynomials.

## Different Interpretations of Operating with Reflexive Polynomials

Consider the polynomial $\Omega = T + Tx + Tx^2 + Tx^3$. Using formal operations we can rewrite it as

$$\Omega = T\omega = T(1 + x)^3.$$

A polynomial, in the detailed form that registers the system's state, represents the process of its development from an external observer's point of view. The same polynomial can be written in two other ways:

$$T + (T + Tx + Tx^2]x = T + [T(1 + x)^2]x.$$

Now $X$ is in the position of the external observer. We can interpret the content of $X$'s mental world in two ways. The left part represents $X$ seeing the system's state, and the right part the dynamics of the system's development. Also, the difference may be explained in terms of the convenience of investigation for the external observer. Then,

$$\Omega = T + [T(1 + x)^2]x$$

signifies the "rolling up" of $X$'s detailed picture performed by the observer.

$X, Y, Z$ do not possess reflexive analysis. For this reason, when we attribute a mental world described by our polynomial to a person, there is a danger that we will make the person contemplate our artificial apparatus and not the content we wished to express by it. Consider the polynomial

$$\Omega = T + [T(1 + x)^n]x.$$

How can letter $n$ be interpreted? If we say that $n$ is a fixed number, then the polynomial must be understood in accord with the comment above. If, however, from $X$'s position, $n$ is an arbitrary number, this means that $X$ has discovered the recursive principle of generating the states he is capable of. How does this situation appear from the position of the external observer, who knows the language of polynomials? After having reflected person $X$, the external observer in his language must register that $n$ is a variable from $X$'s viewpoint (!). Can the external observer continue using the formal principles of the calculus? After making simple transformations, he will obtain

$$T + [T(1 + x)^n]x = T(1 + x)^m, \quad m = n + 1,$$

where $m$ is an integer, but already from the observer's point of view. Doesn't this mean that with this transformation the observer has lost the fact that $X$ discovered the recursive principle, because notation $\Omega = T(1+x)^m$ means only that $X$ is such that operator $\omega = 1+x$ can be used an arbitrary number of times in succession? Yes, the observer has lost this fact, but he can introduce an additional axiom that $X$ knows the principle of induction, which allows him to discover the principle of his own recursive structure.

With any fixed $m$ the polynomial can be represented as

$$\Omega_m = T(1 + x)^m = T + \left[ \sum_{i=1}^{m} T(1 + x)^{i-1} \right] x$$

$$= T + [T + \Omega_1 + \Omega_2 + ... + \Omega_{m-1}]x,$$

where $\Omega_1, \Omega_2, ... \Omega_{m-1}$ constitute a sequence of $X$'s states.

The axiom "allows" $X$ to conduct analysis of his own history, but representation of the states necessary for the analysis is provided by the formal apparatus. The use of the axiom of attributing the principle of induction to $X$ is a kind of step backward to more trivial ways of reasoning.

It is possible to use reasoning of different kind: equation $T+[T(1+x)^n]x = T(1+x)^m$ holds because such is the nature of the processes under consideration. Thus, the possibility of obtaining a generalized picture of the self does not require using the principle of induction. This principle can be considered as a manifestation of the work of deep algebraic processes.

Similar considerations hold for the following situation:

$$\Omega = T(1 + x + y)^m.$$
$$T(1 + x + y)^m = T + [T(1 + x + y)^{m-1}]x + [T(1 + x + y)^{m-1}]y$$
$$= T + [T(1 + x + y)^n]x + [T(1 + x + y)^n]y.$$

Each person can adequately reflect not only the self but also the system of which the self is one element. Understanding the principle or, in the observer's language, using the operator of awareness, does not lead to its replacement. The operator keeps working automatically.

Let person $X$ with operator $\omega=1+x+yx$ discover the principle of domination (not the fact that a given state is dominated by $Y$, but the principle). This principle exists in $X$'s mental world and, as before, is reflected by $Y$. It is depicted in Figure 11 and analytically as

$$T + [T(1 + x + yx)^n + T(1 + x + yx)^n y]x.$$

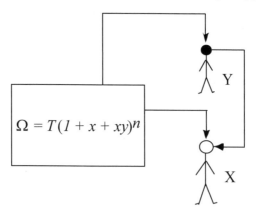

Fig. 11

From the external observer's point of view, discovering the principle does not give $X$ an adequate picture of reality, but rather gives him an adequate picture of the self[3]. 

Let us agree on one more interpretation of $n$. The person may imitate certain situations that, from the observer's point of view, are governed by a definite law, even though the law itself is not known to the person. For example, we may interpret the following notation

$$\Omega = T + [T(1 + x + y)^n]x$$

as the fact that, in $X$'s mental world, a peculiar machine works and drives parameter $n$ along the series of natural integers. This means that the notation registers the dynamics of the process in $X$'s mental world but not the principle.

In the next section, in which we will analyze the Prisoner's Dilemma, the above expression will be understood in just this way.

## Reflexive Polynomials Generating the Prisoner's Dilemma

The Prisoner's Dilemma is an excellent model showing that there are situations in which trivial ideas about rational behavior are not applicable. According to Anatol Rapoport, the Prisoner's Dilemma belongs to the class of paradoxes which "at times appear on the intellectual horizon as harbingers of important scientific and philosophical discoveries" (Rapoport, 1965).

The dilemma is as follows. Two suspects in a crime are arrested and held in separate cells. The prosecutor is convinced that they committed a felony but does not have sufficient evidence to charge them. Each prisoner is given the choice of confessing to the felony or not confessing. If neither confesses, the prosecutor will charge them with a misdemeanor and both will get a light sentence.

---

[3] The representation of a person as $\Omega = T(1 + x + yx)^n$ contains the possibility of $X$'s discovering the principle, because

$$T(1 + x + yx)^n = T + [T(1 + x + yx)^{n-1} + T(1 + x + yx)^{n-1}y]x,$$

which follows from $T(1 + \omega)^n = T + T(1 + \omega)^{n-1}\omega$, when $\omega = x + yx$. An adequate reflection of one's own nature does not eliminate $Y$'s domination.

If both confess, they will receive more serious penalties, but the prosecutor will not ask for the maximum. If the one confesses and the other does not, then the first one will get a significantly lighter sentence, while the other receives the maximum penalty.

The suspects understand that, on the one hand, it is advantageous for both not to confess, but, on the other, it is advantageous for each one to confess if the other does not. Thus, from a rational point of view, neither decision is satisfactory.

We will analyze the reflexive mechanisms generating this dilemma using another example which will facilitate our task. Let two adversaries, $X$ and $Y$, be armed with pistols. If $X$ shoots $Y$, $X$ receives a dollar; if $Y$ shoots $X$, $Y$ receives a dollar. The players are not to be held morally or legally responsible for murder. They make their decisions independently and cannot communicate. What should they do? $X$ is reasoning like this, "Suppose I shoot; I will either gain a dollar or die. If I don't shoot, I certainly won't gain a dollar, but the probability of my death will not be reduced, because my opponent will make his decision independently. My opponent will reason the same way and will pull the trigger. Perhaps, if I don't shoot, he will not shoot either. No, our decisions are not linked. Of course, for both of us, not shooting is advantageous. He will deduce this and not pull the trigger. Aha! Then I'll shoot and gain a dollar! But he will also make the same decision..."

This text shows the reasoning of a player who tries to make a rational decision and faces contradictions at every turn. Both versions are equally unconvincing.

To find the reason for this paradox, imagine that the two gunmen are separated by a partition made from a thin foil mirror that will not stop the bullet. Each of them takes his own mirror image as the image of the opponent. $X$ slowly raises his pistol and sees the opponent with menacing face expression who is raising the pistol. $X$ understands that if he pulls the trigger, the model of the opponent also will pull the trigger.

Since this model is the only means to predict the opponent's behavior, his own shot generates the model's shot. $X$ slowly lowers his pistol; the opponent also lowers the pistol. "I will fool him now," $X$ thinks, "he is using the same model, for sure" and immediately sees the opponent's menacing expression and the warning pistol movement.

In this example and in the preceding one the player uses his own self as a model of his opponent. Any thought of his is also a thought of his opponent's. They stand in front of each other and simultaneously read each other's thoughts. This situation is represented by the following polynomial:

$$\Omega_n = T + (Tx + Ty)x + (Tyx + Txy)x + (Txyx + Tyxy)x + \ldots$$

From $X$'s position, his mental images are also his opponent's mental images. It is impossible to represent the reflexive process preserving such a symmetrical structure within person $X$ using an external multiplier. We have to introduce imbedded operators of awareness. The above polynomial can be rewritten as

$$\Omega_n = T + [T(1 + x + y)^n]x.$$

Independently of the value of $n$, $X$'s mental world is a symmetrical polynomial. Any decision made by $X$ is automatically made by his opponent. If $X$ decides to shoot, the opponent also decides to shoot. If $X$ decides not to shoot, the opponent also decides not to shoot; then $X$ decides to shoot, and his opponent immediately decides the same. We see that the dilemma is generated by the identity of decisions made by the two opponents in $X$'s mental world.

It is important to formulate this problem correctly: who faces the dilemma, the player or the operation researcher, who is supposed to recommend the optimal decision? There is no optimal decision in the Prisoner's Dilemma, but the absence of an optimal decision is not itself a paradox. The paradox appears when the player has his opponent's model and makes the best possible decision, which immediately turns deadly.

Note that if $X$'s structure were different, for example, if $X$ were armed with the operator of awareness

$$\omega = 1 + x + yx,$$

which changes $X$ to

$$\Omega = T + (\Omega + \Omega y)x,$$

no dilemma would appear. $X$ would decide to shoot. If $X$ decides not to shoot, the opponent who reads $X$'s mind will shoot to receive a dollar. If $X$ decides to shoot, the opponent's decision is not clear and there is a chance that he will not shoot, because we do not assume that the opponents follow the principle of "evil for evil."[4]

Therefore, the dilemma appears due to the symmetric structure of the player's mental domain. The Prisoner's Dilemma cannot be solved, but it can be explained.

## Positive and Negative Forms

Consider the following polynomial:

$$\Omega = T + (T + Tx^3)x + (T + Tx + Tx^2 + Ty)y.$$

This is system $\Omega$ from the external observer's position. Let us compare $X$'s and $Y$'s mental world with that of the external observer. To do so we construct the following table:

| External observer | $T$ | $Tx$ | $Tx^4$ | $Ty$ | $Txy$ | $Tx^2y$ | $Ty^2$ | | |
|---|---|---|---|---|---|---|---|---|---|
| $X$ | $T$ | | | | | | | $Tx^3$ | |
| $Y$ | $T$ | $Tx$ | | $Ty$ | | | | | $Tx^2$ |

The empty cells in the second and third lines correspond to the polynomial's terms that exist in the external observer's mind but are absent in $X$'s and $Y$'s. Yet $X$ and $Y$ have extra terms that are absent for the observer: $Tx^3$ at $X$ and $Tx^2$ at $Y$. The terms that are unknown to a person will be represented with a bar above that person letter. In the polynomial above, $Tx$ is unknown to $X$, because his mental image contains only $T$ and $Tx^3$. We will depict this as $Tx\bar{x}$. The

---

[4] In this situation, the operator of awareness $\omega = 1 + x + yx$ results in both players' deaths (if both are "armed" with it), while in the zero-sum game, the same operator generates MaxMin tactics. Thus the same operator may, in different situations, lead to different types of behavior. This fact is very important, because it demonstrates the independence of reflexive processes from the situations where they are involved.

other elements unknown to $X$ are depicted in the same way:

$$Tx^4\bar{x}, \; Ty\bar{x}, \; Txy\bar{x}, \; Tx^2y\bar{x}, \; Ty^2\bar{x} \,,$$

as are the elements unknown to $Y$:

$$Tx^4\bar{y}, \; Txy\bar{y}, \; Tx^2y\bar{y}, \; Ty^2\bar{y} \,.$$

Now we add these terms to the initial polynomial, apply the distributive law and obtain the following:

$$\Omega^* = T + (T + Tx^3)x + (T + Tx + Tx^2 + Ty)y$$
$$+ (Tx^4 + Ty + Txy + Tx^2y + Ty^2)\bar{x}$$
$$+ (Tx^4 + Txy + Tx^2y + Ty^2)\bar{y}.$$

It is easy to see that any finite polynomial $\Omega$ can be represented as

$$\Omega^* = T + \Omega^1 x + \Omega^2 y + \Omega^3 \bar{x} + \Omega^4 \bar{y} \,.$$

This representation allows us to depict not only the content of mental domains, but also the terms that are absent in persons' minds but exist in the system from the external observer's point of view.

The part of $\Omega^*$ which coincides with $\Omega$ will be called its *positive form*, and the sum $\Omega^3\bar{x} + \Omega^4\bar{y}$ will be called its *negative form*.

## Reflexive Polynomial as a Way of Registering Limitations

Imagine the following situation. Every inhabitant of a town, while sitting at their fireplace in the evening, has independently guessed and become absolutely convinced that a circus performance scheduled for the next day will not take place. Then, everyone hears the radio announcement that the performance is cancelled. Did the inhabitants of the town receive new information? At first glance, it seems that they didn't. In reality, however, new information has been obtained. After the announcement, every person in the town knows that every person in the town knows that the circus performance has been cancelled.

Let us designate the town's inhabitants as $e_1$, $e_2$, ..., $e_k$.

Polynomial $\Omega = T + Te_i$ represents each inhabitant at the moment of becoming certain that the circus performance is cancelled. The other inhabitants, with their mental worlds, are not present in each person's mind. Using the negative form, the situation, from the external observer's point of view, is

$$\Omega^* = T + Te_i + \sum_j Te_j \bar{e}_i \quad .$$

The information broadcast by radio removes the bar from $e_i$, and polynomial $\Omega^*$ turns into

$$\Omega^{**} = T + Te_i + \sum_i Te_j e_i = T + \left( T + \sum_j Te_j \right) e_i \ .$$

Thus, we see that the public announcement of information known to everybody changes everyone's reflexive polynomial: it now contains new terms representing the mental domains of other people along with the information they have received.

Reflexive analysis does not allow us to analyze the process of decision making as such, but it sets frameworks which pick out the type of information that may be used in making decisions. When we consider each inhabitant prior to the moment of hearing the information by radio, the only limitation that we have is the absence of $Te_j$ in the inhabitant's mental domain; he himself knows but does not realize that others may also know. The radio announcement moves the person to the other state: the terms $Te_j$ have appeared in his mental domain.

Let $X$ be depicted by the following polynomial:

$$\Omega = T + (T + Tx)x,$$

or, in positive-negative form,

$$\Omega^* = T + (T + Tx)x + Txx\bar{x} \ .$$

Term $Txx\bar{x}$ represents the fact that $Txx$ is unknown to $X$ (but known to the external observer), and $X$ cannot use it for conscious decision making. Person $X$ is free only within the bounds of his mental world $(T + Tx)$.

Suppose that $X$ performs an act of awareness with operator $(1+x)$:

$$[T + (T + Tx)x](1 + x) = T + (T + T + Tx + Txx)x.$$

The previous restriction is now removed: $Txx$ is known to the person, but $Txxx$ is unknown. Thus, the restrictions did not disappear, but they have changed.

In the context of changed restrictions, let us consider locking-in operators. As has been shown, these operators, while changing polynomials, leave some of their very important peculiarities unchanged. For example, when we apply operator $1 + x + yx$ to polynomials of the type $T + (\Omega + \Omega y)x$, the resulting polynomial has the same structure. Thus, the structure represented by polynomial $T + (\Omega + \Omega y)x$ is invariant to the use of operator $1 + x + yx$. We may consider this structure as a restriction of a higher level than those expressed by specific polynomials. Therefore, a *closing-in operator does not remove some structural restrictions*, but changes those imposed by the given operator. A person with only a locking-in operator is "locked" into a class of polynomials of a specific structure. Only a change of the operator of awareness may "free" him and let him leave this class of polynomials.

We can now generalize the concept of awareness. An act of awareness is a procedure that changes restrictions. In this sense, any meaningful function defined in the set of reflexive polynomials and drawing values from the same set can be considered as a special operator of awareness. Still, we need to extend the term "awareness" to include simplifying transformations of polynomials. This leads to strengthening the restrictions. A person loses part of his freedom instead of extending it, as in the case of the operator-multiplier.

## Another Way of Constructing Reflexive Analysis

In the first edition of this book, the operator of awareness was introduced differently. An arbitrary polynomial expressing the relation between two persons was written as $\Omega = T + \Omega_1 x + \Omega_2 y$. Awareness was understood as a reflection of the whole situation by one of the participants. For example, let $X$ perform an act of awareness. The system has changed; polynomial $\Omega$ is "inside" $X$, but $Y$ and the arena $T$ remain unchanged:

$$(T + \Omega_1 x + \Omega_2 y)x + \Omega_2 y + T.$$

This procedure resembles finding an indefinite integral and is designated as

$$\int^{x} \Omega = \Omega x + C, \quad C = \Omega_2 y + T,$$

$$\int^{y} \Omega = \Omega y + C, \quad C = \Omega_1 x + T.$$

The constant $C$ contains the terms in which the outmost right letter is not designating the person performing awareness. When both persons perform awareness simultaneously:

$$\int^{x} \int^{y} \Omega = \Omega x + \Omega y + T.$$

The reverse operation - finding a partial derivative - was also introduced in the first edition. It was interpreted in two ways. On the one hand, it was understood as selecting a person's mental world, and, on the other, as finding the system's previous state (of course, under the condition that the current state was generated by an act of awareness, in the above sense). Formally, the operation of differentiation was defined as follows:

$$\frac{\partial \Omega}{\partial x} = \Omega_1, \quad \frac{\partial \Omega}{\partial y} = \Omega_2.$$

If a polynomial is represented as $\Omega_1 = T + \Omega_3 x + \Omega_4 y$, it is possible to find its second derivative, i.e., to select a person's mental world inside the mental world that was selected earlier:

$$\frac{\partial^2 \Omega}{\partial x \, \partial x} = \Omega_3, \quad \frac{\partial^2 \Omega}{\partial y \, \partial y} = \Omega_4.$$

The procedure of differentiation may be repeated many times until the derivative becomes equal to $T$.

It is easy to see that this way of defining the operator of awareness leads to a small class of polynomials. To expand it, some additional artificial methods have been introduced.

In this edition, the author found it more efficient to use the procedure of multiplying polynomials as an analogue to the process of awareness. The operation of differentiation may be used in the new version of reflexive analysis, but it is interpreted only as a procedure of selecting a person's mental world.

# Chapter II.

# Focal Points and Reflexive Polynomials

Let us conduct a hypothetical experiment. A prisoner is locked in a cell, whose plan is shown in Figure 12, and outside is his partner, who wants to help the prisoner escape.

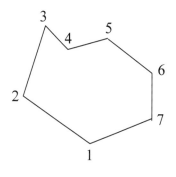

Fig. 12

One person cannot break through the wall by himself, but if they work from opposite sides at the same time, an opening can be created. The wall can be broken only in the corners: 1, 2, 3, 4, 5, 6, 7. How will reasonable persons behave, if prior to starting their work or in the process of it neither the prisoner nor his partner knows how the other will act? The task would be much simpler if one of the corners were thinner than the others, but what if all corners have the same thickness? A simple experiment will show that the choice falls on corner 4. Why? How can two systems coincide without communication? Note that systems without reflexion cannot meet in such situations, because a decision made by either of them is unconnected with the decision made by the other. The meeting in corner 4 is not an accident for reflexive systems which reproduce one another's mental worlds.

For us, since we are reflexive systems, it is obvious that corner

4 must be chosen because it is "strange[1]." Why this tendency toward what is "strange?" Problems of this type concerning meetings without preliminary agreement and informational contacts were investigated earlier by Thomas Shelling (1960). He was the first to prove that the meeting takes place in the strangest places, which he called "focal points." Shelling cited many interesting examples of focal points, but the logical-psychological mechanism of this phenomenon has never been fully explained. It would seem that a reflexive chain "I think that he thinks that I think..." explains the appearance of a focal point. But this chain can explain only the cases in which the relation of preference is given. For example, if during the rain two people want to meet in the park, where there is a garden house, and it is raining, then, such reasoning may help: "I think that he thinks that I think that the garden house is the best place to keep out of the rain." Nevertheless, this kind of reasoning cannot help in the story with a prisoner; the "strange corner" does not have any objective advantage (or subjective preference based, for example, on custom or habit). In this case, there is no relation of preference for a person other than the existence of the other person. A chain of the type "I think that he thinks ..." cannot be completed with a rational justification of the choice, but reflexive analysis allows us to explain the reasons for the focal points' appearance, because it takes account of more complicated structures.

The following polynomial corresponds to the structure "I think that he thinks ...":

$$T + \{T + [T + (T + Ty)]x\}y \ .$$

Inside Y, there is X, and inside X is Y. The depth of enclosure can be arbitrary. For this kind of simple structure, it is expedient to use a special depiction; for example, the above polynomial can be replaced by the following expression:

$$\overrightarrow{YXY},$$

which is read, "Y thinks that X thinks that Y thinks" or "Y knows

---

[1]This obviousness recalls the obviousness in the antic thinking: a stone falls because it is heavy.

that $X$ knows that $Y$ knows." The arrow indicates the order of reading.

These structures are of two classes. Structures of the first class end with the name of the person who conducts the reasoning; for example,

$$\overrightarrow{YXYXY}.$$

The number of letters is odd. As the initial one, $Y$ uses his own reasoning, which is imitated by $X$, then imitated by $Y$ etc.

Structures of the second class end with the name of the other person, for example,

$$\overrightarrow{YXYX}.$$

The number of letters is even. As in the first case, $Y$ uses the reasoning of the other person.

To characterize the depth of imitation in these structures we introduce a parameter of *rank of reflexion* (Lefebvre, 1965a), which is the number of consecutive enclosures. A nested doll may serve as a good illustration: the number of dolls inserted in the largest one corresponds to the rank of reflexion.

The analogy of the nested doll may be used for any polynomial: inside a large doll, there is one, or more than one, smaller dolls, inside each of which are still smaller ones, etc. The number of the inserted dolls is arbitrary. If we put each person in correspondence with dolls of a specific color, the analogy will be complete.

Let us return to the operator of awareness $\omega = 1 + x + yx$ and to the polynomials formed by it:

$$\Omega = T + (\Omega + \Omega y)x.$$

We already know that in a situation of conflict this operator generates the MaxMin strategy, and that in the situation of the prisoner's dilemma, which we analyzed with the example of shooters, this operator generates a shot.

Suppose that a prisoner inside the cell depicted in Figure 12 is equipped with the operator $\omega = 1 + x + yx$. From this prisoner's point of view, his partner imitates any thought of his. We have to distinguish between a decision and its realization. The decision is an element in the person's mental world. The realization of the decision is a component of the person's behavior. Reasoning that justifies a choice of one alternative must include this feature. Reasoning based on frequently met features leads to ambiguous choices; reasoning based on an exclusive feature leads to the unique choice. For example, if six corners are painted red and one is green, the reasoning, "I choose this corner, because it is red," gives six equivalent options, but the reasoning, "I choose the corner because it is green," gives only one option. The informational meaning of features is clear here. If the prisoner "feels" that any thought of his is imitated by the accomplice, then the choice which leads to the minimal number of indistinguishable realizations has the advantage.

The difference in the work of operator $\omega = 1 + x + yx$ in a situation of conflict and in one of the collaboration can be clarified with the following example. $X$ wants to avoid contact with $Y$, while $Y$ wants to encounter $X$. There is a group of black and white points where $X$ may hide (Fig. 13). In his choice, $X$ takes into consideration only color, and the points of the same color are indistinguishable to him.

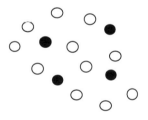

Fig. 13

$X$ can take one of two decisions: he can choose a white point or he can choose a black point. Since, from $X$'s position, $Y$ imitates his every thought, $X$ should choose a white point, so that the probability of $Y$ finding $X$ is lower. Note that the seemingly universal idea of hiding in a featureless element in a subset of many elements is defined by the operator $\omega = 1 + x + yx$. If the person were represented by the polynomial

$$\Omega = T + [T + T(1 + x + yx)^n y]x,$$

he would make the decision to hide in a black point.

Consider now a case where $X$ and $Y$ both want to meet in one of the points depicted in Figure 13. $X$ has only one operator of awareness: $\omega = 1 + x + yx$. Naturally, by virtue of the above explanation, $X$ will choose a black point.

Therefore, the work of operator $\omega = 1 + x + yx$ in the case of solving a joint task under conditions of impossible communication generates the phenomenon of the focal point. (In the given example, a focal set is generated because black points are undistinguishable.) If $Y$ is designed like $X$, i.e., has his own operator of awareness $\omega = 1 + y + xy$, both of them will get to the same "focal set," and if this set consists of one element, they will certainly meet.

Ironically, in the given example, $X$ and $Y$ have inadequate images of each other. In reality, they are such as an external observer sees them.

The operator $\omega = 1 + x + yx$ is not the only one that generates focal points and focal sets. Consider a person represented as

$$\Omega = T + [T\,(1 + x + y)^{n}]x \; .$$

We met this polynomial while analyzing the Prisoner's Dilemma. Such a structure of the mental world can also generate a focal point. Since, for $X$, his partner is identical to him, $X$ believes that the partner will automatically make the same decision as $X$. If $X$ has to choose a point in Figure 13, he will pick a black one, because it follows from the very fact that $X$'s choice is a black point his "mirror partner" will also choose a black point.

So we conclude that the phenomenon of focal points is generated by specific reflexive structures.

Let us construct an example in which a person, on the one hand, generates a focal point, and, on the other, must neutralize deduction. We add a third person, a warden $Z$, to the situation where a prisoner $X$ and his accomplice $Y$ try to break through the wall from opposite sides. $Z$ sits in ambush to watch one of the cell's corners. $X$ and $Y$ know about this, but they don't know which corner he is watching.

Consider the prisoner. To meet the accomplice, $X$ has to choose a focal point. But since, from $X$'s position, this choice will be deduced immediately by the warden, he decides to avoid the focal points; then, however, he loses any chance to break through the wall. A peculiar "fugitive's dilemma" appears due to the operator of awareness $\omega = 1 + x + yx + zx + yzx$, which generates polynomials of the type

$$T + [\Omega + \Omega y + (\Omega + \Omega y)z]x .$$

$(\Omega + \Omega y)$ generates a focal point, and $(\Omega + \Omega y)z$ requires neutralizing an enemy's deduction.

We see that reflexive systems have a reserve of self-organization (absent in systems of other types), which allows them to function expediently without informational exchanges.

There is special interest in the functioning of a system whose information flows are available to its opponent. It is advantageous for the opponent if the system's elements exchange information. On the one hand, this allows the opponent to penetrate the system's intentions, and on the other, it lets him separate hostile elements from neutral ones. The opponent may encourage building confronting coalitions to make his enemies visible, but is powerless if the elements do not exchange information and act synchronically by using the reserve of self-organization inherent in reflexive systems. Although a preliminary agreement on coordinated actions may not be needed, some set of features must be common to all elements; otherwise the elements may generate different focal points (Bongard, 1967).

The situation with a prisoner's cell seems similar to that with cosmic civilizations which do not have contact with each other. When we began searching for cosmic neighbors at a radiation of 21 sm., we assumed that this was a focal point and that the cosmic neighbors "guessed" we would look for their signals there. The focal points play the role of all-seeing eyes.

We are led to propose that reflexive processes represent a universal mechanism that may allow cosmic civilizations to find each other and perform coordinated actions without informational contact.

# Chapter III

# Reflexive Control

In this chapter we will analyze the processes of persons' interactions in conflict. Consider a situation of conflict taking place in the framework of the reflexive polynomial

$$\Omega = T + Tx + (T + Tx)y .$$

The reality that Y sees is not only a picture of the objective arena but also X's image of this arena. We will assume that with this structure, Y understands his adversary's goal and means of problem-solving or doctrine; in this case, Y may set himself the task of manipulating X's decision-making process. Such manipulation is not a matter of directly imposing one's own will on the adversary, but rather of conveying to X the basis for arriving at a decision predetermined by Y. To do so, Y connects to X's "system of images" and begins influencing X's decision-making process (Fig. 14).

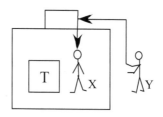

Fig. 14

The process of conveying the basis for decision-making from one person to another will be called *reflexive control* (Lefebvre, 1965, 1966, 1967). Note that this definition suggests only the simplest cases of the phenomenon. Any deceitful movements, provocations, intrigues, camouflage, or, generally, a falsehood of any type constitutes reflexive control. A lie may be complex in structure: for example, conveying truthful information to the adversary, with the intention that the adversary considers it false and makes a decision advantageous to the other.

## Interpreting Reflexive Control As a Special Way to Receive Information about a Partner

How can $X$ obtain the information possessed by $Y$? If $X$ can intercept the channel by which $Z$ sends information to $Y$, then the information will reach $X$, who can then insert it into his model of $Y$ (Fig. 15). $X$ may also connect to the channel by which $Y$ sends information to $Z$ (Fig. 16); finally, it is possible that $Y$ may just provide $X$ the information at his disposal (Fig. 17).

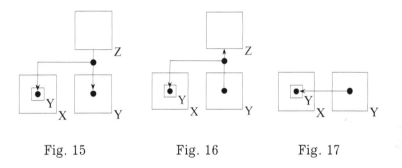

Fig. 15          Fig. 16          Fig. 17

In addition to these "natural" ways of obtaining information, there is one more: $X$ can give $Y$ specially prepared information and simultaneously place this information in his model of $Y$ (Fig. 18). *In this way, $X$ obtains information about $Y$ that $X$ himself has put it there.*

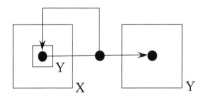

Fig. 18

Reflexive control is a tool used to obtain information about a partner. This information can be of any kind: about the arena, about the self, about the partner, about one's point of view concerning a partner's point of view, etc. The important thing is that, after providing the information, $X$ becomes the possessor of new information about his partner.

We use the term *reflexive control* for conveying the basis on which a particular decision can be made. This is a special means for obtaining information relevant to a partner's behavior[1]. It is only one technique among many. The partner's mental world may be of interest to $X$ for reasons other than simply of influencing his behavior.

Because the simplest types of reflexive control involve manipulation of decisions, we need to highlight the elements of decision-making processes and, at least roughly, establish the connections between them.

## Depicting Simplest Decision-Making Process

Suppose a person is represented by the polynomial $\Omega = T + Tx$. We will expand the meaning of $Tx$ and, in addition to depicting an arena (which we will designate as $A$), include the operative elements necessary for decision-making. The choice of elements depends on which problems we are solving and what details we need.

For example, say we have a space including several distinct locations (element $A$); $X$'s goal is to deliver goods to the locations by truck in a single trip. The arena is mapped on $X$'s tablet, constituting the element $Ax$. By "tablet" we mean the entire system of symbolic means which record the "objective situation." The mapping may be done with varying degrees of precision: some points may be omitted or misplaced, and some nonexistent points may be mapped. From now on $X$ will work not with $A$, but with $Ax$, map his decision on $Ax$, and only after that transfer it to the real arena with greater or lesser success.

---

[1]In Chapter V, we will show how the principle of reflexive control can be realized in simple automata.

$X$ has a goal, $Gx$: to deliver goods from a warehouse to several points by truck in a single trip. To make decisions leading to achievement of his goal, $X$ must work with the tablet. We assume that $X$ possesses a method for solving such problems; we call it his *doctrine* and designate as $Dx$. It may be linear programming or casting dice. In the given example, the doctrine consists of sorting through versions of itineraries to find the shortest one, which is then mapped on the tablet and represents $X$'s decision, $Rx$.

The procedure of decision-making can be represented as follows:

1. The goal is mapped on the tablet: $\dfrac{Gx}{Ax}$,

2. The doctrine is applied to the goal mapped on the tablet:

$$\frac{Gx}{Ax} Dx,$$

3. The result is the decision, $Rx$, mapped on the tablet, $Ax$:

$$\frac{Gx}{Ax} Dx \rightarrow \frac{Rx}{Ax}.$$

Suppose that $X$ has an adversary $Y$, represented by polynomial $T + (T + Tx)y$. Then, the entire situation is:

$$\Omega = T + Tx + (T + Tx)y.$$

Consider the process of $Y$'s decision-making with the intention of intercepting $X$'s truck. The ambush can only be accomplished near point B (located in the forest), but $Y$ needs to know, from which point $X$ will drive toward point B. $Y$ does not have information $X$'s real itinerary, and to make his decision he needs to imitate $X$'s reasoning and infer $X$'s decision. To do so, he uses the following procedure:

$$\frac{Gx}{Ax} Dx \rightarrow \frac{Rx}{Ax},$$

but with a significant difference: $Y$ does not possess $Ax$; $Y$ possesses something that can be called "$Ax$, from $Y$'s point of view." This is a secondary image, $Axy$; it may be different from $Ax$. $Y$ also does not possess $X$'s goal, $Gx$, or $X$'s doctrine, $Dx$. $Y$ possesses "$Gx$, from $Y$'s point of view," $Gxy$, and "$Dx$, from $Y$'s point of view," $Dxy$. The procedure for $Y$'s imitation of $X$'s decision-making is as follows:

$$\frac{Gxy}{Axy} Dxy \rightarrow \frac{Rxy}{Axy}.$$

Suppose $Y$ possesses the knowledge that $X$'s doctrine consists of sorting through different itineraries to find the optimal version ($Y$ knows that $X$ has the appropriate software). Suppose also that $Y$ depicts the area differently from $X$ ($Y$ knows $X$'s depictions from stolen documents) and believes that his own depiction is right. While imitating $X$'s decision-making procedure, $Y$ operates not with the picture on his own tablet but with the one which, from his own point of view, $X$ has on his tablet. After $\frac{Rxy}{Axy}$ is obtained, $Y$ has to transfer this decision to his own tablet:

$$\frac{Rxy}{Axy} \longrightarrow \frac{Rxy}{Ay}$$

Now $Y$ has to map his goal on his tablet, apply his doctrine, which consists of marking the place on $X$'s itinerary where, from $Y$'s point of view, $X$ will approach point B. That is the place where $Y$ will set up his ambush. $Y$ receives his own decision mapped on his own tablet:

$$\frac{Rxy}{Ay} \longrightarrow \frac{RxyGy}{Ay} Dy \longrightarrow \frac{Ry}{Ay}$$

By combining the two expressions into one, we obtain a generalized symbolic depiction of the decision-making procedure for the given situation:

$$\frac{Gxy}{Axy} Dxy \rightarrow \frac{Rxy}{Axy} \rightarrow \frac{Rxy}{Ay} \rightarrow \frac{RxyGy}{Ay} Dy \rightarrow \frac{Ry}{Ay}$$

Analysis of the above example demonstrates that $X$'s pursuit of the optimal result may lead to his defeat, simply because this kind of reasoning is so easy to imitate.

Let us depict $X$'s decision-making procedure for the case where $X$ is presented with polynomial $T + [T + (T + Tx)y]x$. In order to make his decision, $X$ has to imitate $Y$'s decision-making procedure depicted above.

$X$ does not have the elements $Axy$, $Gxy$, $Dxy$ that $Y$ has, but rather "$Axy$, from $X$'s point of view," $Axyx$, "$Gxy$, from $X$'s point of view," "$Gxyx$, and "$Dxy$, from $X$'s point of view," $Dxyx$. The decision-making procedure will be depicted as

$$\frac{Gxyx}{Axyx}Dxyx \rightarrow\!\!\!> \frac{Rxyx}{Axyx} \rightarrow\!\!\!> \frac{Rxyx}{Ayx} \rightarrow\!\!\!> \frac{RxyxGyx}{Ayx}Dyx \rightarrow\!\!\!>$$

$$\rightarrow\!\!\!> \frac{Ryx}{Ayx} \rightarrow\!\!\!> \frac{Ryx}{Ax} \rightarrow\!\!\!> \frac{RyxGx}{Ax}Dx \rightarrow\!\!\!> \frac{Rx}{Ax}$$

We can see in this expression a general law governing construction of "formulas" for polynomials of the given type.

Relationships between people can be much more complicated than this. So far, we have used only polynomials of the type "$X$ thinks that $Y$ thinks that $X$ thinks ..."

Let us note two types of images depicted by the polynomial $\Omega = T + Tx$. In the first type, a person does not include the self in the image of the arena. We may depict this as

$$\Omega = T + \dot{T}x \,,$$

where the dot over $T$ indicates that $X$ excludes the self from his mental representation of the arena (Fig. 19).

In this case, there is a difference between $Ax$, on the one side, and $Dx$ and $Gx$, on the other. $Gx$ is a peculiar function of the reflected relation of the self, as an acting participant, to the arena. Since this type cannot pick out this relation, the goal cannot be reflected. It appears as a peculiar "intention." In other words, the awareness of one's goal as precisely "one's own goal" is possible only given awareness of "one's own actions" or of "one's own relation" to the object. This awareness turns an "intention" into a goal. The concept of a "goal" itself includes the idea of "conscious intention." The goal appears as a specific reflexive formation in teleological constructions. It is meaningless to speak of a bee's goal or an ant's goal.

In the second type, $X$ includes his own "body" and his own actions in the image of the arena (Fig. 20).

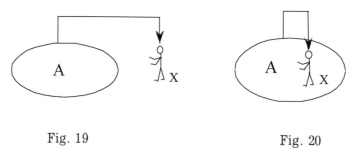

Fig. 19                              Fig. 20

In this case $X$ can include his relation to the object, and the "intention" turns into a goal. Doctrine is a more complicated matter. The possibility of conscious doctrine assumes that someone, in his mental world, distinguishes his image of reality from reality itself. The minimal polynomial in which that may happen is $T + (T + Tx)x$.

Therefore, it is inaccurate to assume that if there are no $Txx$ and $Tyy$ in the polynomials, persons may still have $Dx$ and $Dy$. The author leaves this inaccuracy for the sake of simplifying the explanation.

### Reflexive Control in Conflicts in the Framework of the Polynomial $\Omega = T + Tx + (T + Tx)y$

In such a conflict reflexive control is generally presented as the transformation

$$Txy \rightarrow Tx.$$

$Txy$ here is not a reflected element, but a planned one. This fact must be taken into consideration when marking time:

$$Tx_{i+j}y_i \rightarrow Tx_{i+j}.$$

At moment $i$, Y sees how $Tx$ will look at moment $i+j$, so $Tx_{i+j}y_i$ is the "future" planned in the "present." Henceforth, we will simplify narration by omitting the time subscripts.

In a conflict described with the given polynomial, reflexive control may be conducted by at least one of the following transformations:

$$Axy \rightarrow Ax,$$

$$Gxy \rightarrow Gx,$$

$$Dxy \rightarrow Dx.$$

### Reflexive Control by Picturing Arena $Axy \rightarrow Ax$

This is one of the most common types of control. For example, camouflaging objects is a way of communicating to the adversary that "there is nothing here." Construction of false objects is another technique of this kind.

In many real conflicts it is impossible to convey a complete picture of the arena to an adversary. Instead, a system of reference frames is sent, on the basis of which the adversary constructs his own picture. The construction is a logical procedure, and the one conducting the reflexive control assumes that the adversary is able to make the appropriate inferences. For example, in the second millennium BC, Gideon used torches as a means of reflexive control over his enemy, Medes. The rules of that time required one trumpeter and one torch-bearer for every hundred warriors. Gideon expected Medes to know these rules and to know arithmetic; thus, he provided each of his three hundred soldiers with a porch and a trumpet, assuming that Medes would calculate 300 times 100 and conclude that Gideon had 30,000 troops, leading them to avoid a battle (Medes fled).

Using the reference frame provided by Gideon (torches and trumpets), Medes deduced their picture of the arena:

| Gideon | | Medes |
|--------|--------|--------|
| $Axy$ | $\rightarrow$ | $Ax$ |

## Reflexive Control by Manipulation of the Adversary's Goal $Gxy \rightarrow Gx$

The example of this type is provocation, such as "ideological subversion," treacherous "friendly advice," etc. Another example is the well-known trick where a bank note is placed on the road (attached to an invisible string) and used to provoke passers-by to try to seize it.

## Reflexive control by manipulation of the adversary's doctrine $Dxy \rightarrow Dx$

A doctrine is an operational tool used to arrive at a decision based on a goal and an arena. In the simplest case, doctrine is a system of elementary rules, such as "if $\alpha > \beta$, choose $\alpha$," etc. An opponent's doctrine might be formed by training. For example, if a forward in a soccer game deliberately makes the same mistake several times in a row, the defender develops an automatic response. At the decisive moment, the forward hits the ball correctly, and his team wins.

## Reflexive Control Through the Transformation $Axyx \rightarrow Axy$

This type of control consists of giving the adversary an allegedly correct version of one's own picture of the arena, possibly by leaking false documents. This type also includes leading the adversary to believe that camouflaged objects have not been discovered (although they have) or that false objects are perceived as real ones (when their falsity is known).

## Reflexive Control Through the Transformation $Gxyx \rightarrow Gxy$

An example of this type of reflexive control is a basketball player's feint to the left to make his opponent believe he will move to the left, then moving to the right.

## Reflexive Control Through the Transformation $Dxyx \rightarrow Dxy$

Consider a conflict: $X$ is a pursuer armed with a pistol, and $Y$ is the pursued. $Y$ runs into a cave with six exits (Fig. 21). $X$ can hit $Y$ only if he enters at a point from which he can see the exit chosen by $Y$. Each arrow shows a possible line of fire (Fig. 22).

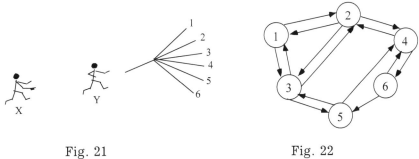

Fig. 21                    Fig. 22

Suppose $X$ informs $Y$ that he will choose an exit by casting dice. If $Y$'s doctrine consists in finding the exit where his likelihood of being hit is minimal and the pursuer's choice of an exit is random, $Y$ chooses exit 6, which vulnerable from only one point (exit 4). All other exits can be hit from at least two points. In reality, however, the pursuer $X$ is not going to cast dice. $X$ reasons that because his adversary thinks that he will cast dice, $Y$ will necessarily choose exit 6. $X$ takes exit 4 and wins.

There are a significant number of kinds of reflexive control. For example, say $X$ and $Y$ are two opposing armies. Many of $X$'s military operations have two purposes: on the one hand, the purely military purpose of inflicting damage on the enemy, and, on the other, that of composing a "text" legible to the adversary, $Y$, leading him to draw false conclusions concerning $X$'s goals.

For example, $X$ concentrates his artillery not in order to fire, but to make $Y$ believe that he is going to fire. In any conflict, significant resources are spent creating "texts" addressed to the adversary.

Sometimes $X$ cannot avoid the transformation $Gx \rightarrow Gxy$, causing $Y$ deduce $X$'s goal from $Axy$. In this situation, in order to hide his true goal, $X$ may choose a goal such that the arena generated by its realization, known to Y, suggests several equally probable goals with the true goal hidden among them:

$$Gx \rightarrow \begin{bmatrix} G^1xy \\ G^2xy \\ \cdots \\ G^nxy \end{bmatrix}$$

      An example of an operation aimed at neutralization of the enemy's intelligence is the German breakthrough of the French front lines on May 15, 1940. The German tanks broke through the French lines in the center, allowing them to move in any direction. At the outset, the French did not know whether the Germans intended to move toward Paris or toward the English Channel. Although it seemed that the Germans were heading west toward the Channel, the French worried that at any moment they might turn South toward Paris. This situation put the French in difficulty.

      The Germans (X) could not hide their actual tank movements from the French (Y), i.e., the transformation $Ax \rightarrow Axy$ had to take place, but the $Ax$ chosen led to two equally probable goals:

$$\text{Movement toward Paris} \rightarrow \begin{bmatrix} \text{1) toward Paris} \\ \text{2) toward the English Channel} \end{bmatrix}$$

In the framework of the polynomial

$$\Omega = T + (T + Tx)y + [T + (T + Tx)y]x,$$

Y might attempt reflexive control by performing one of the following transformations:

$$Axy \rightarrow Ax,$$
$$Gxy \rightarrow Gx,$$
$$Dxy \rightarrow Dx.$$

But since Y's adversary, X, imitates Y's mental world and can assume the possibility of reflexive control, Y's attempt may fail. Suppose that

Y is certain that he has conducted reflexive control successfully. From his position, Y gives a picture of arena, goal, and doctrine to his adversary and thereby has X's mental world at his disposal. In arriving at his decision, Y uses the elements $Axy$, $Gxy$, $Dxy$. In fact, this is a failure of reflexive control. Y has given to X all the elements of his decision, thus simplifying X's task. Instead of the transformations planned by Y, the following ones take place:

$$Axy \rightarrow Axyx,$$
$$Gxy \rightarrow Gxyx,$$
$$Dxy \rightarrow Dxyx.$$

X has obtained information of great importance and can reconstruct a picture of his own self from the adversary's position. Successful reflexive control by X would consist of convincing Y that the transformations which Y planned really did take place. In addition, X may perform the following transformations:

$$Ayx \rightarrow Ay,$$
$$Gyx \rightarrow Gy,$$
$$Dyx \rightarrow Dy.$$

A complete schema of the mutual communication is as follows:

| X | $Gx,\ Dx,\ Ax,\ Gyx,\ Dyx,\ Ayx,\ Gxyx,\ Dxyx,\ Axyx$ |
|---|---|
|   | $\downarrow\ \ \downarrow\ \ \downarrow\ \ \ \uparrow\ \ \ \uparrow\ \ \ \uparrow$ |
| Y | $Gy,\ \ Dy,\ \ Ay,\ \ Gxy,\ \ Dxy,\ \ Axy$ |

In the case where Y does not make an attempt at reflexive control discovered by X, there are no arrows pointing up, and X must develop additional channels of reflexive control and reconstruct his schema as follows:

| X | $Gx,\ Dx,\ Ax,\ Gyx,\ Dyx,\ Ayx,\ Gxyx,\ Dxyx,\ Axyx$ |
|---|---|
|   | $\downarrow\ \ \downarrow\ \ \downarrow\ \ \ \downarrow\ \ \ \downarrow\ \ \ \downarrow$ |
| Y | $Gy,\ \ Dy,\ \ Ay,\ \ Gxy,\ \ Dxy,\ \ Axy$ |

Thus, the "failure" of reflexive control is a special way of transmitting valuable information to the adversary.

Note that in the framework of the polynomial

$$\Omega = T + (T + Tx)y + [T + (T + Tx)y]x,$$

adversaries $X$ and $Y$ conduct reflexive control over each other without being aware of it. $Y$ may try to perform the transformation $Txy \rightarrow Tx$, but his mental world does not contain element $Txy$, and thus he cannot imitate a transformation containing this element. Similarly, $X$ cannot imitate transformation $Txyx \rightarrow Txy$ because there is no element $Txyx$ in $X$'s mental world. This element exists only from the external observer's position. Therefore, in the framework of the given polynomials, individuals are not aware of the reflexive control they conduct[2].

The minimal polynomial in the framework of which the above transformation can be planned is as follows:

$$\Omega = T + [T + Tx + Txy]y + [T + Ty + Tyx + Txy + Txyx]x.$$

By using arrows instead of some "+" to register a possible transformation, as suggested by Baranov, we obtain:

$$\Omega = T + [T + (Tx \leftarrow Txy)]y + [T + (Ty \leftarrow Tyx) + (Txy \leftarrow Txyx)]x.$$

In the above representation we see clearly the deliberate planning of reflexive control.

It seems that the awareness of reflexive control is not a necessary condition of its execution. For this reason, we consider simpler polynomials, in which the schemes of reflexive control are not conscious. The more complicated schemes used in more complicated polynomials can be analyzed in a similar fashion.

## Maneuvering

A special class is constituted by schemes of reflexive control extended in time. In some cases, one adversary transmits his "pseudo

---

[2]V. E. Lepsky and P.V. Baranov drew the author's attention to this fact.

history" to the other, in order that the other extrapolate from this pseudo history to create a plausible (from his point of view) prognosis of future events and make decisions on that basis. One example is a sharp change of *modus operandi* when the adversary knows the old mode very well. Some types of such control have been studied experimentally. The results are presented in Chapter V.

## Artificially Formed Reflexive Structures and Operators of Awareness

The most advanced means of control consists of creating the adversary's reflexive structures. The easiest way to do this is by "insertion" of a well-defined polynomial. When $X$ informs $Y$ that $Z$ is interested in $Y$'s views on the state of affairs in the arena $A$, $X$ forms the following polynomial:

$$\Omega = T + (T + Ty + Tyz)y.$$

This polynomial determines a class of possible decisions by $Y$. In essence, $X$ predetermines the form into which the subsequent information will be poured. Let us emphasize that previously we analyzed reflexive control as a means of influence on the decision-making process by assuming that the one who conducts reflexive control knows the reflexive polynomial that represents his partner. The type of reflexive control that we describe now is focused on the polynomial itself. After it has been created, $X$ can conduct the "usual" reflexive control.

The more interesting type of reflexive control is by forming an operator of awareness. This is a mode of influence directly on the mental domain. This type of control does not envisage a specific goal in a specific situation. If such control is successful, a person becomes enclosed in a narrow class of polynomials and his decisions in completely different situations can be predicted with reasonable assurance by the one who is conducting the reflexive control.

This analysis is not prescriptive. We have chosen and analyzed certain real types of reflexive interactions. Where the discussion seems normative, this is for purposes of emphasis.

# Chapter IV

# Control over Processes of Reflexive Control

Consider the polynomial

$$\Omega = T + (T + Tx)y + [T + (T + Tx)y]x.$$

$X$ adequately reflects $Y$'s reflexive structure. We have already seen that the structure of the above polynomial allows $Y$ to attempt reflexive control over $X$. To simplify discussion, we will denote $X$ as $A$ and $Y$ as $B$, i.e., $B$ can conduct reflexive control over $A$, and $A$ may also conduct reflexive control over $B$, forming his goal, doctrine, etc. In addition, $A$ has a new capability: *controlling the process of B's reflexive control.* There are various reasons for this: for example, to obtain information about $B$'s image of $A$, which would allow $A$ better to predict $B$'s decisions and to achieve his own goals more successfully.

In this chapter we analyze the processes of controlling reflexive control. The analysis is conducted for an arbitrary number of people and an arbitrary number of hierarchies of control. To do this, we develop a special algebraic language allowing us to make complicated processes of this kind plainly visible and to confirm the validity of our schema for controlling reflexive control of an arbitrary complexity.

## Graphical Representation of Controlling the Processes of Reflexive Control

The simplest case of control conducted by individual $A$ over individual $B$, who does not conduct reflexive control, is represented by Figure 23: a straight arrow points from $A$ to $B$. If $B$ becomes involved and begins controlling the process of $A$'s reflexive control, the schema in Figure 23 changes to the one in Figure 24: a curved arrow from $B$ intercepts the arrow between $A$ and $B$. $A$ is conducting reflexive control, and *B* is controlling this control.

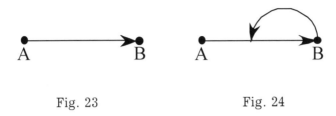

Fig. 23                                      Fig. 24

The next step is easy. *A* reflects the fact that his control is being controlled and connects to the secondary control constructed by *B* (Fig. 25). Schemas of this type can be generalized. If *B* reflects a new reality, *B* may begin constructing control of a higher degree (Fig. 26).

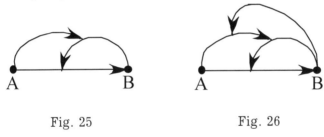

Fig. 25                                      Fig. 26

The schemas in Figure 27 show a special case of reflexive control: *A* is governing reflexive control that he himself has constructed (Fig. 27*a*). Such schemas can be of interest for analyzing the case where *A* is not an individual but rather a complex hierarchical system, in which control over the lower echelon is exercised by the higher echelon. Figure 27*b* depicts *A*'s self-governance: the points *A* and *B* merge into one point and the system is analyzed in less detail (Fig. 27*b*).

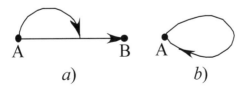

a)                                      b)

Fig. 27

If we are not interested in the hierarchy within $A$, then the schema in Figure 27*a* may be replaced by that in Figure 23. If the hierarchy is essential, we may present $A$ as two elements, $A$ and $C$, and obtain a schema with three individuals; its simplest case is shown in Figure 28: $A$ conducts reflexive control, but it is governed by $C$ (Fig. 28). The case of three individuals becomes more complicated with the appearance of secondary control (Fig. 29). This schema can be presented differently (Fig. 30), but the meaning is the same.

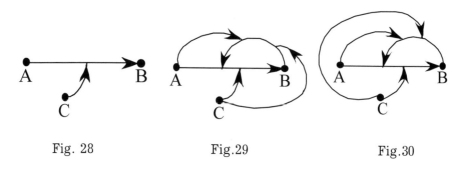

Fig. 28                    Fig.29                         Fig.30

For more complicated cases, simple visual analysis does not allow us to determine the *topological equivalency* of different schemas or to notice fine differences between them. In dealing with ordinary graphs, we can generate corresponding matrices with zeros and ones. The task of determining the topological equivalency of such graphs is reduced to that of comparing the matrices.

In essence, the method described below allows us to determine the equivalency or non-equivalency of various schemas by using elementary algebraic forms. It will also let us draw conclusions concerning the system as a whole.

## Symbolic Representation of Processes of Controlling Reflexive Control

Consider individuals $A$ and $B$ who do not interact. It is a vacuous case. The system consists of two unconnected elements. We represent this system as the sum $A + B$ (Fig. 31).

The system presented in Fig. 23 by a vector from point $A$ to point *B* is depicted as *AB*, and the entire system is *A + B + AB* (Fig. 32).

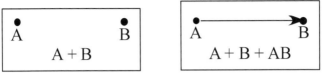

Fig. 31 Fig. 32

Consider the scheme in Figure 24. The curved arrow going from *B* to the arrow *AB* will be designated as *B(AB)* or just *BAB* (Fig. 33). The new arrow that appears in Figure 25 will be designated *A(BAB)* or *ABAB* (Fig. 34).

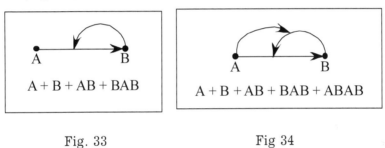

Fig. 33 Fig 34

The rule of constructing symbolic expressions is very simple: every new arrow that ends at another arrow adds the name of the dot where it originated to the left of the name of the arrow where it ends.

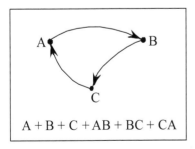

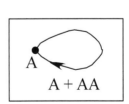

Fig. 35 Fig. 36

*Chapter IV*

The interaction of three individuals (*A* governs *B*, *B* governs *C*, and *C* governs *A*) is depicted in Figure 35, and for the scheme in Figure 27*b*, symbolic representation is given in Figure 36. Here is an example of a more complicated structure (Fig. 37):

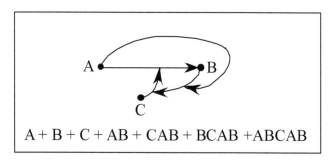

Fig. 37

   To illustrate the use of such representations for purposes of analysis, consider the following situation: *A* wants to send some information to *B* with the goal of conducting reflexive control. *A* can do this only through *C* who, as a rule, deliberately distorts messages, i.e., governs the process of *A*'s reflexive control over *B*. If *B* knows that the information is being distorted, *B* reprimands *C* in a form given to him as a hint by *A*. Thus, *B* governs *C*'s controlling, but *B*'s governance is controlled by *A*. This situation is described by the scheme and polynomial in Fig. 37. We will designate the polynomials shown here as Γ.

   This tool may be useful for the analysis of complex schemas of controlling reflexive control, and especially for determining the equivalency of different graphical representations. The latter are a convenient intermediate schematization of a process under investigation that is difficult to analyze without special techniques.

   Symbolic representation allows us to evaluate to the role of each individual in a structure. The arrows in Figure 37 belong to the subsequent tiers. We will add "weight" to each arrow: *AB* has weight 1; *C(AB)* has weight 2 because it dominates *AB*; *B(CAB)* has weight 3 because it dominates *CAB*, etc. The arrow of every subsequent tier has a weight one unit greater. It is possible to compute the total weight for each arrow originating from a given point, and so to evaluate the role of the corresponding individual in the system, by analyzing the polynomial which describes them. For example, consider polynomial

$$\Gamma = A + B + C + AB + CAB + BCAB + ABCAB.$$

It is natural to assume that $A$, $B$, and $C$ each have a weight equal to zero.

To calculate the total weight for each individual, we add up the number of letters standing to the right of the leftmost letter depicting this individual. For example, individual $A$ is the leftmost twice: in $AB$ (weight 1) and $ABCAB$ (weight 4), so $A$'s total weight is $P(A) = 5$. Similarly, we find $P(B) = 3$, $P(C) = 2$. These numbers characterize the qualitative role of the individuals in the system as a whole.

We can also introduce a qualitative characterization of the reflexive control conducted by individuals over each other. To do so, we calculate a "degree" of dominance. For example, $AB$ is interpreted as $A$'s dominance over $B$ with weight 1, and $BCAB$ is $B$'s dominance over $C$ with weight 1, over $A$ with weight 2, and $B$ over $B$ with weight 3. With repeated appearances, a letter is calculated as many times as the letter appears: $ABCAB$ is interpreted as $A$'s dominance over $B$ with weight 1 and with weight 4, so the total dominance is 5. We can now compose a matrix of relations showing the individuals' influences on one another:

|   | A | B | C |
|---|---|---|---|
| A | 3 | 2 | 1 |
| B | 6 | 3 | 2 |
| C | 2 | 1 | 0 |

We calculate the dominance of each member of the polynomial and sum the result for each individual. Note that dominance over the self signifies a qualitative characteristic of the controlling influences of the others.

A simple case of self-dominance is shown in Figure 33. The schema corresponds to the following polynomial:

$$\Gamma = A + B + AB + BAB,$$

and matrix:

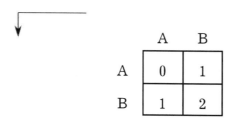

|   | A | B |
|---|---|---|
| A | 0 | 1 |
| B | 1 | 2 |

The analysis of this matrix shows that $A$ and $B$ dominate each other with weight 1, and that $B$'s self-influence exceeds the influence $A$ has over $B$.

Of course, this analysis gives us only the rough qualitative characteristics of individuals' dominance, because the scale is chosen arbitrarily.

## Relations Between $\Gamma$-Polynomials and $\Omega$-Polynomials

Consider polynomial

$$\Omega_1 = T + Tx + (T + Tx)y.$$

In the framework of this polynomial, only $Y$ can conduct reflexive control. Recall that $X$'s other name is $A$, and $Y$'s other name is $B$, and put the following $\Gamma$-polynomial in correspondence with this $\Omega$-polynomial:

$$\Gamma(\Omega_1) = A + B + BA.$$

Consider a more complicated example:

$$\Omega_2 = T + Tx + (T + Tx)y + [T + Tx + (T + Tx)y]z.$$

In the framework of this polynomial, $X$ cannot conduct reflexive control, but $Y$ can reflexively control $X$ by accomplishing the transformation:

$$Txy \rightarrow Tx.$$

Individual $Z$ can reflexively control both $X$ and $Y$ with the transformations

$$Txz \rightarrow Tx,$$
$$(T + Tx)yz \rightarrow (T + Tx)y,$$

i.e., $Z$ can construct $X$'s and $Y$'s mental worlds, such that $Y$ must conduct reflexive control over $X$ preset by $Z$. Let $Z$'s other name be $C$. Then,

$$\Gamma(\Omega_2) = A + B + C + BA + CA + CB + CBA.$$

Consider another example:

$$\Omega_3 = T + (T + Tx)y + (T + Ty)x.$$

In this case, both $X$ and $Y$ can conduct reflexive control:

$$Txy \rightarrow Tx,$$
$$Tyx \rightarrow Ty.$$

Polynomial $\Omega_3$ corresponds to the following $\Gamma$-polynomial:

$$\Gamma(\Omega_3) = A + B + AB + BA.$$

Two more examples arc $\Omega_4$ and $\Omega_5$.

$$\Omega_4 = T + Tyx + Txy.$$

$X$ and $Y$ have the same structure. From $X$'s point of view, $Y$ has an image of an arena, while $X$ believes that, in reality, there is no arena at all. $X$ may try to influence $Y$'s image of the arena, but $Y$ does not have an image of the arena, he has only the image of $X$'s image of the arena. $Y$ also believes that, in reality, there is no arena. Thus, $X$'s attempt to place the arena image in $Y$'s mental world as well as $Y$'s attempt to place the arena image in $X$'s mental world will fail. In the framework of polynomial $\Omega_4$ the transformations

$$Txy \rightarrow Tx,$$
$$Tyx \rightarrow Ty$$

cannot take place. Therefore, since reflexive control is impossible,

$$\Gamma(\Omega_4) = A + B.$$

Consider a system represented by polynomial $\Omega_5$:

$$\Omega_5 = T + (T + Tx)y.$$

Individual $A$ is absent, but from $B$'s point of view he exists. $B$ may imitate reflexive control, but from an external observer's point of view, it is undirected. Thus, $\Omega_5$ corresponds to

$$G(\Omega_5) = B.$$

Let us assume that it is not necessary for an individual to have an adequate image of his adversary's mental world in order to govern the process of reflexive control. Consider the polynomial

$$\Omega = T + Tx + (T + Tx + Txy)y + Txyz.$$

We will assume not only that $Z$ can conduct reflexive control over $Y$ by performing transformation

$$Txyz \rightarrow Txy,$$

but also that $Z$ can govern $Y$'s reflexive control over $X$, i.e., $Z$ can influence the transformation

$$Txy \rightarrow Tx.$$

Of course, concerning this way of controlling reflexive control, we cannot say that it is conscious. In fact, we are noting only the possibility of such influence.

It is possible to formulate a general rule for constructing the polynomial $\Gamma(\Omega)$ corresponding to polynomial $\Omega$. To do so, we introduce a concept of *majorization* among the monomials of a polynomial $\Omega$. A monomial $Ta_1, a_2, \ldots, a_k, a_{k+1}$ is a majorant in relation to $Ta_1, a_2, \ldots, a_k$, where $a_i$ is the name of an individual.

Represent our polynomial $\Omega$ as a graph, whose nodes are monomials and whose arrows indicate majorization. If an arrow points from $A$ to $B$, this means that $A$ majorizes $B$ (Fig. 38). Every monomial has the name of the individual to whom it belongs. Only one arrow may originate at a node, because any monomial can majorize only one other monomial. Now we introduce a concept of *itinerary*.

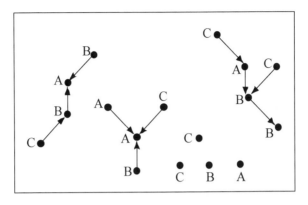

Fig. 38

Consider any two nodes, α and β. By moving along arrows from α we may or may not reach β. If it is possible to reach α from β, we say that α and β are connected by an itinerary. The itinerary, connecting two dots, is unique. We designate each itinerary by the names of its nodes in order of movement, including the beginning and the end. Then, we find the set of all itineraries and construct a list of all designations in such a way that each designation appears only once; combine them with symbol "+," add the names of individuals, and receive the desired polynomial Γ(Ω). Note that there is no unique solution for the inverse problem. For an arbitrary Γ-polynomial there are infinite number of corresponding Ω-polynomials.

Polynomial Ω, representing the interaction of two individuals, can be written as

$$\Omega = T + \Omega'x + \Omega''y.$$

An external observer can construct Γ(Ω); individuals X and Y can construct Γ(Ω') and Γ(Ω"), respectively. It is interesting that there exist polynomials Ω such that

$$\Gamma(\Omega) = \Gamma(\Omega') = \Gamma(\Omega'').$$

For example,
$$\Omega = T + \underbrace{(Ty + Tyx)}_{A+B+AB}x + \underbrace{(T + Ty^2 + Ty^2x)}_{A+B+AB}y.$$
$$\underbrace{\phantom{\Omega = T + (Ty + Tyx)x + (T + Ty^2 + Ty^2x)y}}_{A+B+AB}$$

# Chapter V

# The Devices that Turn Fears into Reality

It is very difficult to study reflexive control in a real conflict; it is more efficient to construct special automata that simulate various schemas of reflexive control. We name them *dribbling*. These automata allow us to investigate the objective characteristics of human reflexion. In addition, it turns out that we can construct an automaton with a paradoxical peculiarity: it optimizes its work on the basis of human counteraction.

Before describing the experiments, let us consider the following situation. There is a town with a labyrinth of streets that cross at intersections. A traveler is in the center of the town and wants to leave it. The traveler does not know where exits are, does not remember what streets he has already taken, and does not recognize intersections when he sees them for the second time. At each intersection, the traveler asks for directions to the nearest exit. The inhabitants are hostile to the traveler, however, and conspire against him to keep him in the town as long as possible.

An experiment conducted by the author (Lefebvre, 1969) has demonstrated that if a simple automaton plays the role of the traveler and a human subject plays the role of the town's inhabitants, the traveler will get out of the labyrinth faster than by randomly wandering without paying attention to the intentionally misleading directions.

### Functioning of a System under Conditions of Human Counteraction. Description of the Experiment

The system consists of three units (Fig. 39). The first unit, the board, is a labyrinth with two little lamps at each node: one green and one yellow. Five nodes around the circumference are exits. A human subject is given the task of keeping a "traveler," whose movements are shown by yellow lights, from leaving the maze.

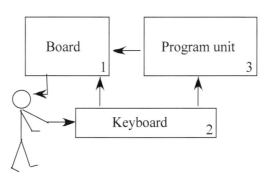

Fig. 39

The traveler does not know where the exits are, nor does he remember the nodes he has visited. The traveler moves only after the subject gives him directions from the keyboard (unit 2). The traveler can move only from one node to an adjacent node. The subject sees the direction he gives to the traveler as a green light. The direction is input to the program unit that controls the traveler's movements.

The controlling program is based on the following principle. At each node the traveler may either obey or disobey the subject. In the case of obeying, the traveler moves to the node indicated by the subject; in the case of disobeying, the traveler moves to the node opposite the indicated one. A table of opposing nodes is stored in the program unit[1]. The schedule of the traveler's movements can be represented as a sequence of integers with alternating '+' and '-'. We tested the following schedule:

+5 -6 +2 -4 +4 -1 +1 -2 +4 -3 +2 -1 +1 -3 +4 -3 +4 -2 +1 -1
+3 -2 +3 -4 +2 -1 +5 -3.

The '+' before a number means going to the indicated node, '-' means going to the opposite node. The absolute value of the integer

---

[1]The table of opposing nodes is given at the end of this chapter.

indicates the number of times a traveler obeys or disobeys directions. This schedule was obtained experimentally and did not change during the control sessions. The problem to be solved by the device is moving the traveler from the central node to one of the exits. The labyrinth is given in Figure 40. At the beginning the traveler is in node 13; nodes 1, 5, 9, 24 and 26 are exits.

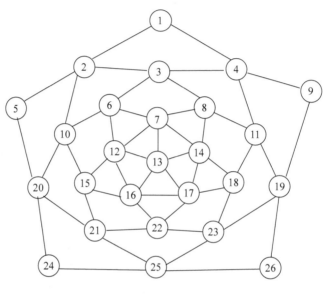

Fig. 40

The procedure of the experiment is as follows. A subject sits in front of the board; near him is the keyboard. The experimenter gives the instructions:

> In front of you is a maze. A dot-traveler is here (the yellow light flashes in node 13). The dot can move along the lines connecting the nodes (the dot moves from node 13 to an adjacent node and goes back). The dot has the task of leaving the maze. The exits are painted red. The dot does not know where the exits are; in addition, it has no memory and does not remember nodes it has visited. You can give direction using the green light (a green flash appears in an adjacent node). Your task is to give directions such that the dot remains inside the maze as long as possible. If you succeed in keeping it for 25 moves, you are the winner. Otherwise, the dot wins. Think of the dot as a human being who wants to get out of the maze, and try not to let it escape.

When the subject asks how the dot reacts to directions, the experimenter says that he does not know, that the reaction program is hidden in the device, and that, in principle, the dot behaves as it wishes. Then the game begins. The subjects are not limited by time. The experimenter logs each game by writing down the number of the node where the green light flashes and the number of the node where the yellow light moves. I have to mention that in this experiment, unit 3 was not automatic. The experimenter's assistant had the table of opposing nodes (p. 82) and used it to switch on the yellow light in the appropriate node.

The group of subjects consisted of 32 students from the Moscow Institute of Energy. Each subject played two games with the device. Each game continued until either the subject or the machine won. All games were recorded. The following tables show the number of games in which the traveler reached the exit in the indicated number of moves.

Table 1. First Game Results

| Number of moves | 7 | 8 | 9 | 10 | 11 | 15 | 16 | 17 | 25 | 37 | 39 | 46 |
|---|---|---|---|---|---|---|---|---|---|---|---|---|
| Number of games | 0 | 4 | 6 | 5 | 4 | 4 | 4 | 1 | 1 | 1 | 1 | 1 |

Table 2. Second Game Results

| Number of moves | 7 | 8 | 9 | 10 | 11 | 12 | 16 | 17 | 19 | 27 | 28 | 29 | 39 | 52 | 56 |
|---|---|---|---|---|---|---|---|---|---|---|---|---|---|---|---|
| Number of games | 1 | 6 | 8 | 2 | 2 | 1 | 2 | 2 | 1 | 1 | 1 | 1 | 1 | 1 | 1 |

According to these data, the mean numbers of the traveler's moves were 15 for the first game and 18 for the second. In addition, a distribution function was constructed:

$$\rho(m) = \frac{k(m)}{n},$$

where $n$ is the number of games, and $k(m)$ is the number of games lower than $m$.

## Functioning of a System Without Human Counteraction

**A model imitating the system functioning.** The functioning of a system without counteraction was simulated by the computer. The model imitated a game in which the system works according to the given schedule and the opponent's directions were equally probable for every adjacent node at every move.

This model can be considered as movement without counteraction when, at every node, the traveler casts dice, and depending on the results, either follows the given direction or moves to the opposite node. Since the nodes are not opposed one to one, the use of the strategy described may change the mean number of moves as compared with simple movement without the concept of "opposite." In our case, the "opposite" does not favor the traveler. By casting dice the traveler increases the number of his moves in the maze to 27, while for random movement it is 25.

To prove the optimization, we have to compare the work of the system with counteraction (directions given by a human) with its work in the absence of human interference (the system uses the dice casting and the same schedule as in the game with the subject).

**A function of distribution for random movement.** Let $\rho_0(m)$ be the probability that the game ends within a number of moves not exceeding $m$. In our case, $\rho_0(m)$ can be found by representing the movements as Markov chain (Fig. 41). The first element in the chain corresponds to the central node, 13 (see Fig. 40); the second element to the level consisting of nodes 7, 12, 14, 16, 17; the third element to the level consisting of nodes 6, 8, 22, 15, 18; the fourth element to the level of nodes 3, 10, 11, 21, 23; the fifth one to the level of nodes 2, 4, 19, 20, 25; the sixth element corresponds to the points of absorption 1, 5, 9, 24, 26.

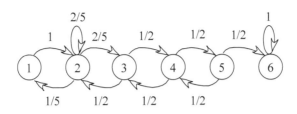

Fig. 41

The matrix for this chain is as follows:

$$A = \begin{Vmatrix} 0 & 1 & 0 & 0 & 0 & 0 \\ 0.2 & 0.4 & 0.4 & 0 & 0 & 0 \\ 0 & 0.5 & 0 & 0.5 & 0 & 0 \\ 0 & 0 & 0.5 & 0 & 0.5 & 0 \\ 0 & 0 & 0 & 0.5 & 0 & 0.5 \\ 0 & 0 & 0 & 0 & 0 & 1 \end{Vmatrix}$$

According to Markov chain theory, the probability that a dot is absorbed in a number of moves not exceeding $m$ is equal to element $a_{16}$ of matrix $\Lambda^m$, where $m$ is an index of the exponent of the power to which the matrix must be raised.

## Comparison of the System under Conditions of Counteraction and in the Absence of Counteraction. Discussion

The mean number of the traveler's moves in the absence of counteraction was 25. The mean number of the traveler's moves under conditions of human counteraction was 15 in the first game and 18 in the second. This data allows us to conclude that the system optimizes its performance as a consequence of human counteraction.

A general picture of the system's functioning is illustrated by the functions of distribution in Fig. 42, where the graph I shows $\rho(m)$ in random wandering; graph II shows $\rho(m)$ in the first games, and graph III is $\rho(m)$ in the second games. As an additional criterion of optimization, the difference between medians can be chosen (the median is the value of $m = m_0$ such that $\rho(m_0)=1/2$). The median in random wandering is 19, in the first games 11, and in the second games 10. The median shift to the left that appeared under conditions of counteraction can be considered as a sign of optimization.

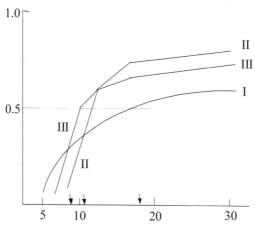

Fig. 42

**Graphical recording of a game**.

Table 3. A single game records.

| Number of the level indicated by a subject | Number of the node in which a green light flashes | Number ofthe node in which a yellow light flashes | Number of the level indicated by a subject |
|---|---|---|---|
| 2 | 7 | 7 | 2 |
| 2 | 14 | 14 | 2 |
| 1 | 13 | 13 | 1 |
| 2 | 16 | 16 | 2 |
| 2 | 12 | 12 | 2 |
| 1 | 13 | 15 | 3 |
| 2 | 16 | 10 | 4 |
| 3 | 15 | 2 | 5 |
| 4 | 3 | 5 | 6 |

This table can be represented by a special graph (Fig. 43):

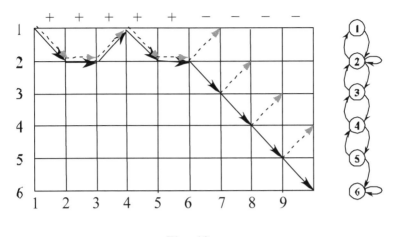

Fig. 43

The horizontal lines correspond to levels. Their relations are given in accordance with Fig. 41. Each vertical line corresponds to a move. Bold arrows indicate the traveler's movement by levels. The result is the traveler's trajectory. Dotted lines show the subject's directions.

**Explanation of the optimization.** The program conducts reflexive control over a human subject leading to optimization. The control proceeds as follows: the first five moves are performed with the symbol "+" (dotted arrows in Fig. 43 coincide with bold arrows). A human interprets these moves as the traveler's obedience to instructions. During these five moves the program creates in the subject the belief that it is obedient, i.e., that its doctrine is to obey. After this belief is formed, the traveler begins to exploit it by choosing nodes opposite to those indicated by the subject. The traveler knows the subject's goal and the subject's model of the traveler's doctrine. This is enough for the traveler to see the maze from the subject's point of view. Thus, if the traveler chooses nodes opposite to the directed ones, it will move closer to the exits. Of course, it is not the machine that is reasoning in this way; it is the programmer constructing the sequence of numbers.

Figure 43 shows that, after the first five moves, the subject keeps trying to lead the traveler toward the center of the maze and, in this way, lets it know the directions leading to the exits. It is actually the subject himself who leads the traveler out of the maze.

There were subjects who, after the traveler stopped obeying, understood that the traveler's real doctrine consisted of choosing nodes opposite the indicated ones and began pointing at peripheral nodes with the intention of leading the traveler toward the center. Consider the following game (Fig. 44).

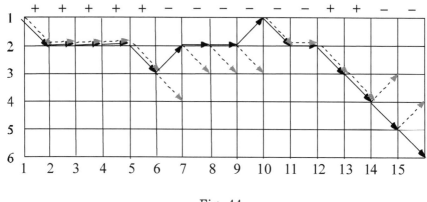

Fig. 44

Believing that the traveler would be obedient, the subject in his sixth move tries to lead him to the fourth level, but the traveler stops obeying and goes to level two. After this, the subject's idea of the traveler's doctrine changes. Now the subject is sure that the traveler is disobedient, so he must point to the periphery. Note that merely understanding that the traveler is disobedient does not automatically change the subject's strategy. The subject must determine which directions should be given in this case (most subjects could not). Starting at move 12, the traveler is obedient once again, but the subject stays with his new method and gives directions leading to the periphery. Since the traveler is now obedient (it takes the subject two moves to realize this), the subject begins pointing at nodes leading to the center. After convincing the subject of its obedience, the traveler again stops obeying (move 14) and exits the maze.

The traveler won by creating a particular behavior in the subject, taking advantage of it, and, at the right moment, changing his behavior abruptly, using this new one and so on, on the average outrunning the subject's ability to adapt to his changes[2].

We deliberately avoid the term "teaching" with reference to forming the subject's behavior. If the subject is aware of the fact that he is being "taught," he may change his behavior accordingly. The concept of reflexive control refers more accurately to the situation: the traveler only provides a basis for decision-making.

The program was selected experimentally, because information concerning how quickly subjects could adjust their awareness was not available in advance. Judging from the results of similar experiments, the rate of awareness is similar for most subjects.

Let us note one important point. There is no feedback between the algorithm controlling the traveler's actions and the subject's actions[3]. The traveler does not know the success or failure of its strategies in opposition to the human subject. The sequence of numbers used by the automaton represents an *a priori* model of the subject performing acts of awareness.

The automaton, if we anthropomorphize it, possesses a model of the subject, including his acts of awareness and consequent changes of the subject's behavior; because of this, the automaton can predict them without feedback. Of course, there was feedback during preliminary experimentation and construction of the "*a priori* model." In any specific experiment, however, the automaton can function without feedback from the subject; in other words, the human subject has no influence over the algorithm dictating obedience or disobedience.

Note also the following. The subject is playing against an algorithm. The traveler is an element of the subject's mental world generated by the instructions given to him. The experimenter's goal is to use the instructions (themselves a special form of reflexive control) to create the necessary "game mental world" in the subject. The yellow light must become the traveler, and the picture of the

---

[2]Note that the table of opposing nodes allows the traveler to remain at the second level after certain directions. Because of this, at the seventh and eights steps, the traveler does not move to the first level. At the tenth step, the traveler disobeys, but since it is at the first level, it must leave that level.

[3]Refer once more to the schema in Fig. 18, Chapter III.

graph must become the town. The experimenter has no part in the subject's mental image.

It is possible to construct another experiment in which the subject is aware of participating in an artificial situation (more precisely, the awareness of the artificial situation becomes dominant for him): moreover, the subject is aware that he is playing against an inflexible program set up by the experimenter. In this case, the subject's mental world has a completely different structure. The experimenter is now part of the subject's mental world, and the game is between the subject and the experimenter. It is a fundamentally different experiment.

We have conducted several experiments of this type. The subject, who was one of the experimenters, knew that the other experimenter had created a program especially for him. The difference between the two experiments is illustrated by Figure 45 (the schema of the first experiment) and Figure 46 (the schema of the second experiment). In the second case, the subject is aware of the experimental situation.

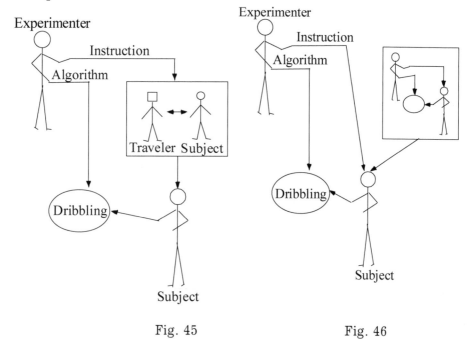

Fig. 45                                      Fig. 46

Experimenters could compete with one another over who could better imitate the program prepared for him by the other. This competition became one component of the process determining who could better imitate the other's mental world.

This procedure is as follows. Both subjects write programs for each other at the same time. Neither is allowed to observe his partner's playing against the program, because, by analyzing the play, he might (we say "intuitively," since we do not know it occurs) reconstruct the program written for him. The competition among a group of subjects leads us to construct a graph of their differing ability to imitate.

## Experiments Conducted by P. V. Baranov and A. F. Trudoliubov

By further elaborating the experiment described above, Baranov and Trudoliubov conducted two more experiments. In the first one the subjects were presented with a symmetrical labyrinth with two exits (nodes 9 and 26); the traveler starts in node 12 (Fig. 47).

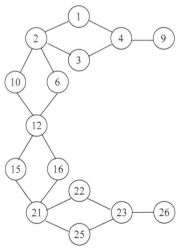

Fig. 47

The subjects' task was to choose one of the exits secretly and prevent the traveler from escaping through that exit. The idea of the algorithm, which again functioned without feedback, was analogous to the one used in the previous experiment: first, the traveler (a program) made the human subject believe that the traveler was obedient, then employed this belief and created a new one, etc. The experiment demonstrated that, about 72% of the time, the traveler escapes through the exit chosen by the subject as forbidden. As in the previous experiment, a yellow light shows the traveler's moves, and the subject's directions are shown by a flashing green light. The algorithm governing the traveler's moves was as follows:

$$+2 \; -3 \; +1 \; -3 \; +1 \; -2 \; +3 \; -1 \; +3 \; -1$$

Recall that "+" means to comply with directions, and "-" means to move to the node opposite to the one specified by the directions. The subjects consisted of 51 male and 10 female students from the Moscow Institute of Energy. The ratios of the automaton's victories to its losses were 38/13 for males and 7/3 for females. The average length of a game was 18 moves; the average length of games won by the automaton was 15 moves; the average length of games lost by the automaton was 26.5 moves; the length of the automaton random moves was $25^4$.

For the second experiment the subjects were given a different task. The experimenter offered each subject a choice of four goals; they were required to keep their choice secret from the experimenter.

1. Lead the traveler toward exit 9 and do not let it escape through exit 26.

2. Lead the traveler toward exit 26 and do not let it escape through exit 9.

3. Lead the traveler out of the labyrinth through any exit as soon as possible, at but in any case in fewer than 25 moves.

4. Keep the traveler inside the labyrinth as long as possible, but in any case for more than 25 moves.

The automaton had one algorithm for all four tasks:

---

[4]This is correct under the condition that the probability of getting to the exit from a node adjacent to it is 1/2, which is justified due to the linear structure of the labyrinth.

+2 -3 +1 -4 +1 -3 +3 -1 +4 -1 +1 -1

The subjects were students of the Moscow Pedagogical Institute. The results are given in Table 4:

Table 4

| | Goal numbers | | | | |
|---|---|---|---|---|---|
| | 1 | 2 | 3 | 4 | Total |
| The number of games | 26 | 12 | 27 | 20 | 85 |
| The ratio of the automaton's victories to losses | 19/7 | 12/0 | 6/27 | 17/3 | 54/31 |

The automaton won for all tasks except 3. The experimenters then corrected the algorithm by changing its conclusion and conducted a new series of experiments. The new algorithm was as follows:

+2 -3 +1 -4 +1 -3 +3 -2 +3 -1 +1 -1

The peculiarity of the new series was that the subjects did not choose the task but rather drew pieces of paper with numbers from a box. In reality, all the papers had the same number 3. Table 5 shows the recalculated results:

Table 5

| | Goal numbers | | | | |
|---|---|---|---|---|---|
| | 1 | 2 | 3 | 4 | Total |
| The number of games | 26 | 12 | 39 | 20 | 97 |
| The ratio of the automaton's victories to losses | 18/8 | 10/2 | 22/17 | 18/2 | 68/29 |

This additional series was conducted in such a way as to make it possible to compare it with the other three tasks from the previous series. This was possible because most of the games ended in the early part of the algorithm, which had not been changed, and those games that ran longer were counted as "lost" by the automaton. The results of it are extremely important. The experiment demonstrated the possibility of creating an effective schema of reflexive control that is independent of the "plot" of the experimental game. Moreover, the tasks differ in their criteria for victory: two of them it is the number of moves, and for the other two it is the choice of one of two alternatives. The experiment also demonstrates that it is possible to find a schema of reflexive control that is not sensitive to criteria. The only important thing for the system is the subject's counteraction; the criterion for the counteraction is secondary.

**Experiment by V. E. Lepsky**

Another type of experiment was conducted by Lepsky. A subject was presented with a scoreboard where he could switch on one of two lamps: green or yellow. There were two more lamps, also green and yellow, for the experimenter to switch (Fig. 48).

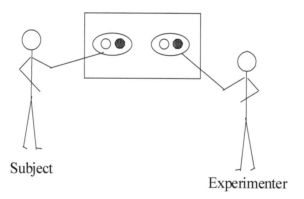

Fig. 48

Without telling anyone, the subject chooses one of two goals: either to turn on his lamp of the same color as the experimenter's or to turn on his lamp of a different color from the experimenter's. This game corresponds to a zero-sum game with the following matrix:

|  | green | yellow |
|---|---|---|
| green | +1, -1 | -1, +1 |
| yellow | -1, +1 | +1, -1 |

The first goal corresponds to choosing a row of the matrix; the second goal corresponds to choosing a column. This matrix is helpful because it allows us to disregard the possibility that the human loses due to deficient skills in choosing the optimal strategy, while the program uses the optimal strategy and wins. In a game with this matrix, if one player uses the optimal strategy, both players win with equal probability. The experimenter did not know which goal and which strategy the subject had chosen; he functioned as an automaton using the fixed sequence of numbers with "+" or "-":

$$+2\ -3\ +1\ -3\ +1\ -2\ +2\ -2\ +4$$

or, in a different notation:

$$+\ +\ -\ -\ -\ +\ -\ -\ -\ +\ -\ -\ +\ +\ -\ -\ +\ +\ +\ +$$

At the beginning of each game the experimenter cast dice to assign "+"and "-" to particular colors.

Thirty subjects participated in this experiment, each playing twenty games. The results demonstrated that program wins with a probability close to 80%. It is important to add that the program succeeds only if there are special instructions giving the subject the desire to defend himself while suppressing the desire to attack.

A comparison of all the experiments shows that certain reflexive invariants reveal the self in human conflicts independently of any particular plot of the particular games, suggesting that reflexive phenomenology should be an object of specialized research.

Automata that optimize their work as a result of human counteraction can be interpreted as "devices which turn fear into reality." Such devices transfer into a certain state either very rarely or have uniform distributions of outcomes; if a human tries to prevent the system from transferring to a certain state, however, it transfers to this very state soon and frequently.

## Table 6. Opposing Nodes[5]

| 2 | 11 | 18 | 3 | 12 | 19 | 4 |
|---|---|---|---|---|---|---|
| 1 -10 | 4 - 18 | 11 - 17 | 2 - 8 | 6 - 13 | 9 - 23 | 1 - 11 |
| 10 - 1 | 18 - 4 | 17 - 11 | 8 - 2 | 7 - 16 | 23 - 9 | 11 - 1 |
| 3 - 5 | 8 - 19 | 14 - 23 | 4 - 6 | 13 - 15 | 11 - 26 | 3 - 9 |
| 5 - 3 | 19 - 8 | 23 - 14 | 6 - 4 | 15 - 13 | 26 - 11 | 9 - 3 |

| 13 | 20 | 6 | 14 | 21 | 7 | 15 |
|---|---|---|---|---|---|---|
| 7 - 17 | 5 - 21 | 2 - 8 | 7 - 17 | 15 - 25 | 14 - 6 | 10 - 16 |
| 12 - 14 | 21 - 5 | 8 - 2 | 8 - 17 | 25 - 15 | 6 - 13 | 16 - 10 |
| 14 - 16 | 14 - 24 | 3 - 12 | 17 - 8 | 20 - 22 | 8 - 13 | 12 - 21 |
| 16 - 7 | 24 - 10 | 12 - 3 | 13 - 18 | 22 - 20 | 12 - 8 | 21 - 12 |
| 17 - 12 | | | 18 - 13 | | | |

| 22 | 8 | 16 | 23 | 10 | 17 | 25 |
|---|---|---|---|---|---|---|
| 16 - 23 | 3 - 14 | 12 - 17 | 18 - 25 | 2 - 15 | 13 - 22 | 21 - 26 |
| 23 - 16 | 14 - 3 | 13 - 22 | 25 - 18 | 15 - 2 | 22 - 13 | 26 - 21 |
| 17 - 21 | 11 - 7 | 15 - 17 | 19 - 22 | 6 - 20 | 14 - 16 | 24 - 23 |
| 21 - 17 | 7 - 11 | 17 - 15 | 22 - 19 | 20 - 6 | 16 - 18 | 23 - 24 |
| | 13 - 6 | 22 - 13 | | | 18 - 16 | |

---

[5]The number above each column is the number of the node where the traveler is; the left number in a column indicates the neighboring node; the right number indicates the opposite node.

# Chapter VI

## Reflexive Connections in Groups

### A Simplest Reflexive Model of a Social Organism

Let $T$ be an arena of material things on which is deployed a reflexive show, and let $e_1, e_2. \ldots , e_n$ be individuals, each of whom has his own mental image of the arena: $Te_1, Te_2. \ldots , Te_n$. In addition, some of the individuals may have mental images of other individuals' mental images: $Te_ie_j$, etc. Thus, a symbolic representation of the reflexive connections is as follows:

$$\Omega = T + \sum_i Te_i + \sum_i \sum_j Te_je_i + \sum_i \sum_j \sum_k Te_ke_je_i + \ldots \qquad (1)$$

This kind of series expansion allows us to see the reflexive characteristics of a social group or part of one. Not every individual has all $e_i$ (some of them may be missing); a representation of a system of "rational creatures," however, must contain at least the third member of this series:

$$\sum_i \sum_j Te_je_i$$

The level of a civilization's development can be characterized, to some degree, by the increasing length of this series. In the near future, space studies will require the construction of models of cosmic civilizations. It seems plausible that the possibility of representing a system with reflexive polynomials is the characteristic that distinguishes civilizations as a special class of systems. Civilizations differ in principle from systems of other types, such as, for example, a colony of cells constituting a living organism or a colony of separate specimens of the anthill-type. A system of the anthill-type is represented as

$$\Omega = T + \sum_i Te_i \, , \qquad (2)$$

where $Te_i$ are models of the environment that allow each "ant" to orient itself. A system grows into a simplest civilization when its series expansion becomes

$$\Omega = T + \sum_i Te_i + \sum_i \sum_j Te_j e_i = T + \sum_i (T + \sum_j Te_j)e_i \, . \qquad (3)$$

Each individual, along with its image of the arena, has images of other individuals' images. Most likely, in the anthill-type systems such secondary images are absent. In systems with quadratic terms, human-like communication and spiritual values may appear. Note that civilizations constitute only one class of systems. Reflexion does not necessarily make a system more adaptive or perfect. We can imagine a community of primitive creatures developed no more than a bee as individuals; if, however, they can adopt mental images from one another, that is, if they can see the world through each other's eyes, we must consider these communities to be civilizations. Reflexive structures are not dependent on the mode of the systems' functioning. For example, using instruments is not necessary for a system's realization of the reflexive polynomial.

Therefore, any civilization and any subsystem of a civilization that has the main features of the whole can be represented as

$$\Omega = T + \sum_i \Omega_i e_i \, , \qquad (4)$$

where $\Omega_i$ is the mental world of individual $e_i$. In this representation, the mental world constitutes a "special civilization"; such modeling allows us to analyze the system's reflexive structure.

It is possible to find "reflexive inequality" in individuals: images of some of them exist in almost everybody's mental world (e.g., movie stars), while others exist only in a few. It is up to us to set the depth of the analysis the diversity of individuals' images. Finally, the representation of a system as a series allows us to see how reflexive control is conducted. In particular, an investigator may set out to bring a system into a certain state. If the investigator is one of the individuals in the system, he must include his own reflexive images in the series. In this case, a "free" member of the series, $T$, is identical to $Te_i$ ($e_i$ being the name of the investigator).

The task of reflexive control involves influencing some members of the series. Further development of this model may allow us to find connections between the free member of the series and other members and thus to describe the system's evolution. In addition, it may become possible to explain the functions of various semiotic systems: religious rites, fashions, etc., through their manifestations in the members of the series.

Note that if element $Te_2e_3e_5$ exists in polynomial $\Omega$, that does not necessarily mean that $Te_2e_3$, i.e., "$e_2$'s image of the arena from $e_3$ point of view" exists there. $Te_2e_3e_5$ might have been invented by another individual, for example, $e_6$, who formed the element $Te_2e_3e_5e_6$ and conducted reflexive control $Te_2e_3e_5e_6 \rightarrow Te_2e_3e_5$.

If some people believe in goblins and even know the goblins' point of view on certain matters, the investigator can depict such reality using a symbolic series: he simply gives a name to the goblin and includes it to the series. If the investigator also believes in goblins, then that "individual" will be part of the series beginning with the second member $\sum_{i}^{n} Te_i$ (goblins' bodies exist in $T$). If the investigator does not believe in goblins, that "individual" will be present only in the following members of the series, and the right letter, $e_i$, will never depict a goblin. Similarly, individuals who have died or fictitious characters such as Othello or Tom Sawyer can also be included in a symbolic series. By taking into account the elements' time dependance, it is possible to include past history and future plans in the system. For example, if $e_\tau$ constructs $Te_\tau$, then its realization means a trivial case of reflexive control: $Te_\tau \rightarrow T$.

Groups can be characterized by the numbers of mutual images corresponding to different members of the series. Simple mathematical models register the presence or absence of a particular image. This allows us to determine reflexively closed groups of individuals, those that do not have images of individuals from other groups, and to divide in this way a social organism into large elements.

It is very interesting to apply reflexive analysis to literature. It may help us to find important features of reflexive structures that operate in social organisms. In addition, it may prove useful for critical analysis.

### Individuals, Positions, Roles, and Gitics

We feel it important to distinguish the concepts of "individual" and "position." An *individual* is an abstraction identifying a person as a primary unit of observation. We give this unit a name and endow it with a mental world. An individual as a social phenomenon is not necessarily holistic. In different situations or social relations the individual may have different mental images generated by different operators of awareness. The individual may assume different guises and play different roles, requiring us to label him differently in different situations.

Therefore, one physical individual may include several positions, each of which exists as an independent social phenomenon. They may be even in conflict with each other. We cannot say that the individual "chooses" one or another position; rather the position "inhabits" him. The individual has no higher principle to govern the interrelations among positions. Only external influences force him to adopt one or another position. The positional boundaries do not lie between individuals. They exist in their mental worlds, separating one position from another, sometimes by an unbridgeable abyss.

The notion of "role" has acquired significant meaning in contemporary sociology and social psychology. This concept allows us to distinguish canonical social positions from the concrete individuals occupying them. This, in turn, allows us to identify different structures in a social organism. The individual may "play different roles" in the family, at work, in an informal group, etc. Obviously, this notion implies an analogy of theatrical drama to social process. In this case, normative interrelations (natural or legal) constitute the framework for dramatic action.

Despite the usefulness of this concept, it can lead to confusion. The primary use of the concept "role" is to designate the individual's position in the functional structure of the social organism. For example, in his family, he is the boss, at work an ordinary employee, among friends a scapegoat. The notion of a role also implies artificiality. The role is something played. In this case, we have to do with a special procedure of constructing one's own image (for oneself and for others). To play a role means to conduct reflexive control. Sometimes the role is externally imposed. An example is the role of a prisoner, who among other things must address his guard in a prescribed manner. In essence, it is a borderline case of reflexive control phasing into adaptation: the prisoner must manage his guard's perceptions in such a way as to avoid punishment. Thus, to "play a role" always means to conduct reflexive control. Both applications of the concept "role" contain a hidden assumption about the possibility that the individual can change his role by a conscious decision.

If we now extend the traditional phenomenology to include "mental worlds" and processes of awareness as fully qualified elements, the concept of a role becomes too weak. A different concept is needed.

For example, the religious individual does not play the role of a believer (if we ignore certain subtle aspects of ritual). He is a believer in fact. He cannot change this state, because the phenomenon of religious faith implies a very particular "mental display" (see Chapter I). God is an element of the mental world that cannot be dislodged by processes of awareness. Indeed, the believer may play the role of unbeliever, but he cannot become an unbeliever merely by conscious decision.

Therefore, when we pass from the analysis of social structures to the study of the mental world as a legitimate element of the social organism, we need a new concept to express features of the individual that are invariant to the process of awareness, i.e., are inherent to that particular individual. One such feature is the mental display; for several independent positions, each has its corresponding mental display.

This fundamental character, which the individual cannot alter and which predetermines the structure of his mental world (and through it his behavior), the author calls *gitic*. This concept can be used flexibly, but always as the contrary of what is artificial and self-regulating.

## Reflexive Currency

Recall the chain "$X$ thinks that $Y$ thinks that $X$ thinks ...," which we examined in Chapter II. We can substitute for "thinks" any word from the list "knows does - not know, consider - does not consider, informed - uninformed." These chains express the direction to the mental world, involving both self and other, "I know that I know that I know ...," with many possible variations: "$X$ knows that $Y$ does not know that $X$ knows."

The fact that the concepts *know* and *think* can link together the above chains reveals their reflexive nature. Their function in language is to express processes related to reflexive phenomenology. For example, the concept "go" cannot be used in such chains; the phrase "$X$ goes that $Y$ goes that $X$ goes" has no meaning.

It is possible to include the nuances of tense and mood inherent in natural languages in a meaningful chain: "$X$ knows that $Y$ knew that $X$ would not know." The concepts under analysis reveal individuals' mental worlds with information or in its absence. Other nuances such as "$X$ is sure that $Y$ is sure that $X$ is sure" suggest the quality of information. Those concepts as "is sure is not sure, supposes does not suppose" illustrate the ability of the natural languages to express transitive reflexive chains.

These patterns are of interest mainly to linguists. We are interested in them because they conform to our mathematical model and describe the reality for which we constructed it. Other possible chains exist: "$X$ values that $Y$ values that $X$ values ..." Hear we speak of values, not information, but the patterns manifest the same or a similar reflexive regularity. Just as the other person's information is a component of my information, so the other person's values are components of my values.

In this section, the author presents an illustrative model which describes the formation of values by reflexive individuals. This model was constructed together with P.V. Baranov and V.E. Lepsky (Lefebvre et al., 1969). Consider an individual in a group. The values of the other members of the group to which the individual belongs influence his values. The others' suffering, in some sense, is his suffering. Of course, the level of suffering is different. For some individuals, others' suffering may be felt as a joy.

This activity is regulated by what we will term *internal currency*. The necessity of this concept arises when we analyze so-called irrational acts of social behavior. In some cases, acquiring money or other external value ("official" currency) may damage one's self-respect, i.e., diminish the internal currency. In other cases, the quantity of internal currency depends on the surrounding individuals. Sometimes, causing harm to others increases one's internal currency; at other times, it reduces it. Sometimes an individual may translate the external currency received by other individuals into his own internal currency, or his own internal currency may depend on these other individuals' internal currency.

If we assume that internal currencies can be compared objectively, then it is possible to introduce the notion of a group's internal currency, or even a society's internal currency. In this case, a complicated ethical problem arises: how to choose the coefficients by which individuals' internal currency can be computed into the total currency. Indeed, the very formulation of this problem depends on the investigator's ethical position (Rapoport, 1964).

An approach to the problem of converting external currency into internal currency is described below. The main difficulty is to find measurable parameters that can be included in the polynomial corresponding to an individual. Say $X$ receives payoff $A$, and his partner payoff $B$. $A$ and $B$ are payoffs in external official currency. The method of receiving payoffs is not taken into consideration.

Let us introduce two parameters: $\alpha$ (characterizing $X$'s attitude to the self) and $\beta$ (characterizing $X$'s attitude to his partner (Rapoport, 1956)). We define $X$'s internal currency as

$$H_1^{(X)} = A + A\alpha + B\beta. \tag{5}$$

$Y$ is also characterized by parameters $\alpha$ (characterizing $Y$'s attitude to the self) and $\beta$ (characterizing $Y$'s attitude to his partner). $Y$'s internal currency is

$$H_1^{(Y)} = B + B\alpha + A\beta. \tag{6}$$

The absolute value of $\alpha$ is the coefficient of the amplification of the payoff. For example, by receiving a small payoff in the external

currency a person may gain significant internal currency. The value
of β depends on the person's attitude to his partner. If Y's
tribulations and joys are indifferent to X, β=0. If X empathizes with
Y's feelings and optimizes Y's profits, sometimes to the prejudice of
his own, then β is positive and may be greater than α. If X dislikes
Y (B<0) and Y's gains hurts X's feelings, then β<0, so βB>0; this
means that X's internal currency increases as a result of Y's losses
(Rapoport, 1956).

   As the next step in constructing this model, suppose that X
and Y have become aware of their own internal currency, as if each
of them carried out computations using formulas (5) and (6). Of
course, no such computations exist in reality. We use this feeble
arithmetical analogy in the absence of any other means for dealing
with internal currency.

   We assume that X imitates his adversary's value system and
ascribes the coefficients α and β to him (unconsciously, of course).
They are X's reflexive characteristics. Assume that the internal
currency obtained above is processed in the same way as the official
currency with which we began:

$$H_2^{(X)} = H_1^{(X)} + H_1^{(X)}\alpha + H_1^{(Y)}\beta ,$$

$$H_2^{(Y)} = H_1^{(Y)} + H_1^{(Y)}\alpha + H_1^{(X)}\beta .$$

At every act of awareness, one's own internal currency is multiplied
by α, and the partner's internal currency by β; then both values are
added to the self's internal currency:

$$H_n^{(X)} = H_{n-1}^{(X)} + H_{n-1}^{(X)}\alpha + H_{n-1}^{(Y)}\beta , \tag{7}$$

$$H_n^{(Y)} = H_{n-1}^{(Y)} + H_{n-1}^{(Y)}\alpha + H_{n-1}^{(X)}\beta . \tag{8}$$

The iterative process of generating internal currency resembles the
process of expanding the polynomial $\Omega = T(1 + x + y)^n$. The
difference is that for reflexive polynomials, an act of awareness keeps
the reflexive structure of the preceding polynomial, while for internal
currency, the history is not structure but valuation.

   Let us consider the case where the sequence of iterations

converges to a certain value. To avoid examining a huge number of variants of finite awareness, we will obtain an operator which would allow us to find the internal currency when we know only the value of the official currency and the coefficients $\alpha$ and $\beta$. Define $H_o^{(X)} = A$, $H_o^{(Y)} = B$. To express $H_n^{(X)}$ in terms of $A$, $B$, $\alpha$, and $\beta$, we add terms (7) and (8):

$$H_n^{(X)} + H_n^{(Y)} = \left(H_{n-1}^{(X)} + H_{n-1}^{(Y)}\right) + \left(H_{n-1}^{(X)} + H_{n-1}^{(Y)}\right)\alpha$$
$$+ \left(H_{n-1}^{(X)} + H_{n-1}^{(Y)}\right)\beta = (A + B)(1 + \alpha + \beta)^n. \tag{9}$$

Now subtract (8) from (7):

$$H_n^{(X)} - H_n^{(Y)} = \left(H_{n-1}^{(X)} - H_{n-1}^{(Y)}\right) + \left(H_{n-1}^{(X)} - H_{n-1}^{(Y)}\right)\alpha$$
$$- \left(H_{n-1}^{(X)} - H_{n-1}^{(Y)}\right)\beta = (A - B)(1 + \alpha - \beta)^n. \tag{10}$$

Up to this point we consider parameters $\alpha$ and $\beta$ independent of the rank of reflexion, i.e., from the finite number of acts of awareness. Now we assume that in each situation of a series of situations, a well-defined rank of reflexion manifests itself. We assume also that $X$ and $Y$ interact in a series of situations such that $X$ can realize any finite rank of reflexion. In the situation where $X$ performs one act of awareness, $\alpha = \alpha_0$ and $\beta = \beta_0$. In the situation where $X$'s rank of reflexion is $n^1$,

$$\alpha = \frac{\alpha_0}{n}, \quad \beta = \frac{\beta_0}{n}.$$

Assume that the above values can be used in (9) and (10). We make this assumption because, on the one hand, these values are sufficient for the internal currency sequences to converge, and, on the other, they give us a convenient limit presentation. The more general way of choosing coefficients leads to infinite products. Unfortunately, limit formulas can become quite complicated. Thus, we compromise by

---

[1]We are using a certain trick here, by assuming that the situation predetermines the number of acts of awareness and that the number of acts of awareness predetermines the coefficients.

assuming that this particular coefficient is dependent on the rank of reflexion. In passing to the limit at $n \to \infty$, we obtain the following limit values for sum and difference:

$$H^{(X)} + H^{(Y)} = (A + B)\exp(\alpha_0 + \beta_0), \tag{11}$$

$$H^{(X)} - H^{(Y)} = (A - B)\exp(\alpha_0 - \beta_0). \tag{12}$$

Use (11) and (12) to express $H^{(X)}$ through $A$, $B$, $\alpha_0$, $\beta_0$:

$$H^{(X)} = \frac{1}{2}(A + B)\exp(\alpha_0 + \beta_0) + \frac{1}{2}(A - B)\exp(\alpha_0 - \beta_0) =$$

$$= \left[ A\frac{\exp\beta_0 + \exp(-\beta_0)}{2} + B\frac{\exp\beta_0 - \exp(-\beta_0)}{2} \right]\exp\alpha_0.$$

Since

$$\frac{\exp\beta_0 + \exp(-\beta_0)}{2} = \cosh\beta_0, \quad \frac{\exp\beta_0 - \exp(-\beta_0)}{2} = \sinh\beta_0,$$

we obtain

$$H^{(X)} = (A\cosh\beta_0 + B\sinh\beta_0)\exp\alpha_0. \tag{13}$$

In our idealized case, this limit value allows us to pass from external currency to internal. Parameter $\beta_0$ is interpreted now as "the angle between players"; it is more important than $\alpha_0$ because the latter is responsible only for the scale of internal currency in the matrix of these currencies.

It is possible to find various test games specifically designed to determine the parameter $\beta_0$. Below we describe the idea of one such game. Without loss of generality, we suppose that the maximal payoff in the official currency that each player may receive is equal to 1. In the test, $X$ chooses an arbitrary point on the circle $x^2 + y^2 = 1$ (see Fig. 49). $X$'s official payoff will be denoted $\cos\varphi$, and the adversary's payoff is denoted $\sin\varphi$. Thus, $A = \cos\varphi$, and $B = \sin\varphi$. We will suppose that $X$ chooses an angle such that his internal currency reaches the maximum. By knowing the value of angle $\varphi$ chosen by $X$ and assuming that $X$ tends to maximize his payoff, we can find the value of $\beta_0$ with which the internal currency is maximal. An elementary analysis shows that for any real value of $\beta_0$, the optimal value of $\varphi$ is located in the interval $-\dfrac{\pi}{4} < \varphi < \dfrac{\pi}{4}$.

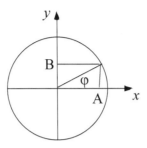

Fig. 49

Therefore, this hypothetical experiment allows us to determine the real parameter $\beta_0$ by angle $\varphi$ or to conclude that $X$ was not trying to solve the problem of optimization (if $|\varphi| \geq \dfrac{\pi}{4}$).

Note that this model does not register the influence of the subject's choice on parameter $\beta_0$, which will be used by his opponent. The point is that, by making a "selfish" decision, $X$ disrupts $Y$'s "humanism." So, $X$'s actual decisions depend on his control over $Y$'s parameter $\beta_0$.

The above scheme can be generalized for an arbitrary number of individuals whose relations are expressed by real numbers. Let $m$ be the number of individuals, and $W_0 = \left\| a_{ij} \right\|$ the matrix of their relations, where $a_{ij}$ is the coefficient of the $i$-th individual to the $j$-th one. An obvious generalization of relations (5) and (6) is the following system:

$$H_n^{(1)} = H_{n-1}^1 + \sum_{j=1}^{m} H_{n-1}^j \frac{\alpha_{1j}}{n};$$

$$\cdots\cdots\cdots\cdots\cdots\cdots\cdots \qquad (14)$$

$$H_n^{(m)} = H_{n-1}^m + \sum_{j=1}^{m} H_{n-1}^j \frac{\alpha_{mj}}{n}.$$

And in matrix form:

$$\begin{pmatrix} H_n^{(1)} \\ H_n^{(m)} \end{pmatrix} = \begin{pmatrix} H_{n-1}^{(1)} \\ H_{n-1}^{(m)} \end{pmatrix} + \frac{1}{n} W_0 \begin{pmatrix} H_{n-1}^{(1)} \\ H_{n-1}^{(m)} \end{pmatrix} = \left( E + \frac{W_0}{n} \right) \begin{pmatrix} H_{n-1}^{(1)} \\ H_{n-1}^{(m)} \end{pmatrix}. \quad (15)$$

By applying $(n\text{-}1)$ times the recursive relation (15), we obtain

$$\begin{pmatrix} H_n^{(1)} \\ H_n^{(m)} \end{pmatrix} = \left( E + \frac{W_0}{n} \right)^n \begin{pmatrix} H_n^{(1)} \\ H_n^{(m)} \end{pmatrix},$$

or, abbreviated,

$$H_n = \left( E + \frac{W_0}{n} \right)^n H_0.$$

By passing to the limit at $n \to \infty$, we obtain the final expression:

$$H = \exp(W_0)H_0. \quad (16)$$

The expression above allows calculation of the internal currency for each individual using the matrix of relations and a column of payoffs. Let us emphasize that this example is purely illustrative, to demonstrate a recursive way of creating internal currency.

### The Origin of Reflexion and Language

The existence of reflexive relationships distinguishes social groups of human beings from systems of other types. Members of a social group imitate each other's reasoning, which serves the function of socialization. By mutual imitation of reasoning, a group may function for a long time without direct informational contacts between its members and still maintain its integrity. The imitation of reasoning serves as a special means of coordination and synchronization for members' activities.

In this section we will consider the example of a primitive social group and investigate the origin of the individual 0-rank reflexion. In this case, the individual's tablet does not contain him.

We will distinguish between a leader and the rank-and-file members of the group. The only function of the leader is to organize the subordinates into appropriate structures such that each of them performs part of the labor, $\tau_1$, $\tau_2$, ... . These procedures are not related internally, only through the leader's communications. The rank-and-file members deal with the reality of work. The leader deals with a special reality, that is, a group of people performing work. The leader does not belong to that group. To perform his function as an organizer, the leader must map this special reality on his tablet, transform it into a project, and fulfill it. The tablet contains the rank-and-file members along with the procedures that they perform. Thus, we begin with two points:

1 - ordinary labor procedures performed by subordinates,

2 - a special labor procedure performed by the leader with respect to a special object, a group of people, with the help of a special semiotic tool, the tablet.

Evidently, we can establish the origin of a human group as the moment when disparate persons are transformed into a self-reflexive system, i.e., with the appearance of semiotic tools for planning the group's work as a whole.

Now we introduce restrictions on the number of rank-and-file members, while leaving the volume of work the same. In a small group, the leader must perform ordinary functions together with his functions as leader, for example, in tasks concerning the distribution of food.

A new line of development begins here. The leader now combines both labor and organizational activities (Fig. 50a). For the first time, individual activity becomes internally organized. A mechanism previously used at the scale of the group is now projected onto individual activity. The leader turns into a self-reflexive system.

Managing the group (command) must be the leader's individual activity, but in the case when the leader is also a member of the group, there is no need for command, because there is no distance, that is, a direct connection appears between the tablet and the leader's own activity. The combination of these two types of activity through their mapping on the tablet signifies the appearance of individual reflexion.

The "ego" originates as an external material substitute for the leader. At first, the loss of this material surrogate means the loss of reflexion. Only then, after being reflected through a physiological apparatus, the material apparatus is reconstituted in the "head." Individual consciousness cannot originate in the "head" To explain the process of its origin, it is necessary to explore the structure of group activity and the evolution of semiotic tools. The problem of human origin as well as of the origin of human groups are, first and foremost, semiotic problems.

Zero-rank reflexion arises in the leader, but the tool which the leader has possesses a peculiar recursive quality. Under certain conditions, the tablet itself and the activity mapped there can be mapped on the same tablet. We see here the special case of planning individual intellectual activity (Fig. 50*b*).

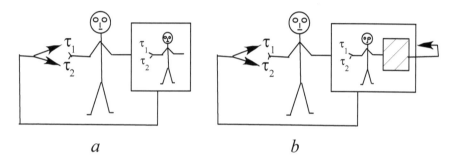

$a$                                      $b$

Fig. 50

When the group appropriates the procedures for imitating its own intellectual activity, new opportunities arise. The imitation by an individual of his own intellectual activity allows the member to imitate the activity of other members without constructing new semiotic tools. Rank-and-file members who have mastered the means of imitation do not depend on instruction from the leader. When direct contact with the leader is impossible to establish, the members can imitate the leader's reasoning, make decisions, convey instructions to themselves, and act in accordance with his decision ("my father would reason this way"). The possession of this mechanism brings with it the potential for reflexive conflicts and leads to the appearance of regulating institutions (religion, ideology, etc.).

It is possible to construct another schema of the origin of individual reflexion. Consider a situation, in which, as in the first schema, the semiotic tools of representing the external world have already made their appearance, but they are external in relation to the acting subject. The process of internalization, i.e., the transformation of external acts of management into "mental functioning," has not yet taken place in human phylogenesis (Vygotsky, 1934; Galperin, 1966; Leontiev, 1965).

The appearance of external semiotic tools, functioning as the surrogates for real objects, allows a "creature" whose behavior is determined by the "receptive field," i.e., by the field of "sensually perceived" objects, to acquire the perception of "sensually non-perceived " objects[2].

In order to have an "image of a river," a rope is needed to serve as perceptual analogue to the river. It is not the "image of the river" that generates the analogy "river - rope," but the use of the rope in a peculiar semantic function in relation to a sensually non-perceived entity, the river, that allows us to compensate this "sensual non-perception." The rope organizes various partial images related to the river and lines them up in a specific sequence. Movement along the "rope" manages their changes.

The entire further intellectual evolution of the social human being is directed toward developing means of representing increasingly wide areas of reality which are impossible to perceive directly. The solar system cannot be perceived as a whole. It is necessary to create a special model as a means to the organization of "social experience" and "social memory."

Let us return to the problem of the origin of reflexion. The "ego" is not perceived sensually. Like the river or the solar system, the "ego" must first appear through an external analogue.

Consider the hypothetical situation of the socialization of two primitive individuals, who possess external semiotic tools for representing the world but do not yet have language. They communicate using a tablet. $X$ perceives $Y$ but does not perceive the self. He registers $Y$'s existence in the model using a pebble, for

---

[2]In this regard, Shchedrovitsky's distinction between "sensually single" and "sensually multiple" is of interest.

example. Similarly, Y perceives X but does not perceive the self. He registers X's existence using another pebble. The tablet is their common domain. It converts them into a unified system. While, separately, each of them had only an image of the other, together they have two images. Thereby, each of them, separately, also becomes a possessor of two images: the self and the other (Fig. 51).

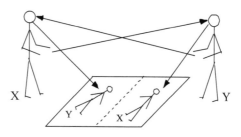

Fig. 51

Thus, we see that it is possible to construct a model of the origin of individual reflexion in cooperative activity-communication on a semiotic tablet. Of course, this is only a preliminary sketch of the model; we have not broached the problem of self-identification.

The above model offers a new approach to the problem of language's origin, the main difficulty of which consists in the impossibility of deducing language from animal systems of communication. Animal "language" is a set of signals directed to "switching-on" distinct actions. It has no capacity for the denotation of an object or phenomenon. In language, we perceive signs as belonging to the realm of things; actions are also registered in language as peculiar things. There have been numerous attempts to overcome this difficulty and deduce language from animal systems of signalization, but this approach seems unpromising.

Let us return to our scheme of communication via a tablet. Let the tablet be a fixed installation. To communicate, one must come to a determined place and assume a certain "working" position. Suppose that the installation must be relocated[3]. It has to be dismantled, and then reassembled in the new place. In order to reconstruct it at the

---

[3]The author appreciates Shchedrovitsky's productive discussion of this topic.

new location, the disassembled parts must be specially marked. The dismantled tablet may be regarded as a linear sequence of its marked parts. In essence, this sequence is already a linguistic expression. Language is in fact a stream of disassembled elements. In the process of phylogenesis, a special sound is assigned to each element of the dismantled system. Thus, audible language is a code denoting not the elements of the real world but the elements of a disassembled "tablet" appropriately marked. The tablet, its disassembling and reassembling, are internalized in the process of phylogenesis. Communication eventually dispenses with the external technical device.

A speech act becomes a tool for transmitting the marked parts of the speaker's dismantled device, and the process of understanding becomes a process of assembling a special configuration on the tablet from those parts. Let us emphasize that the elements of language do not have analogues in the real world. Their analogues are the elements on the tablet which, as a whole, has a direct relation to reality.

A model which represents subjects that communicate without linguistic capacities allows us to pose the problem of communicating with dolphins in a new way[4]. To do so, we would need to construct special tools to enable the dolphin to perceive the self and the experimenter as elements of an external model. Communication would have to take the form of two-way interaction with the model. If, for example, the experimenter wished to meet the dolphin at a specific spot in the pool, he would move the picture of himself to that spot on the model and then move the picture of the dolphin to the same spot. If the experiment were successful, i.e., the dolphin and the experimenter met at the chosen spot, this would be evidence of real contact between two civilizations.

---

[4]The author thanks G.E. Zhuravlev, I.M. Krein, and G.L. Smolyan for interesting discussions on this topic.

# Chapter VII

# Objects as Systems

Every generation of scientists is convinced that their world view has, in principle, a correct grasp of reality; it may not be complete in every detail, but it can be built on and ultimately brought to perfection. There is an assumption that the development of theory follows a special law: each earlier theory must be a particular case of the one that succeeds it. This assumption is based on an idea of approximation, namely, that in a historic sequence of theories, each new theory is a closer approximation to the absolute truth.

This idea of approximation neglects the fact that a scientist's world view is largely determined by the "research templates" at his disposal. The same object can be represented in different ways and, generally speaking, it is pointless to argue over which system representation is correct. The choice of system representation depends on what is convenient for the researcher or the scientific field.

The study of complex objects is a primary task of most scientific disciplines. Each science develops its own specific methods and may represent the objects differently. Nevertheless, scientists usually say that they study systems.

The concept of *system* has penetrated all areas of knowledge. This suggests that there should be special constructs in the field of knowledge itself. One of the main tasks in the study of scientific thought is to describe these constructs in their pure form (Schedrovitsky, 1964).

The concept of *system* is usually linked with other concepts such as *connection*, *structure*, and *element*. In different fields, researchers assign different meanings to these concepts. In all areas, however, the concept of system implies, on the one hand, examination of the object as a whole together with its external parameters and, on the other hand, seeing the object as a complex of elements whose connections constitute its structure (Blauberg et al., 1969).

## Representation of the Object as a System

In most research, the way an object is represented as a system is determined by its initial division into elements, since the connections among elements depend on the character of the elements themselves. For example, in deciphering ancient texts, the separation of elements in the original graphic object is crucial. Only after such separation has been made can one start to investigate connections and structures by comparison of the selected elements. An incorrect initial separation dooms the project to failure.

In all established scientific disciplines, there are traditional methods of analysis that are considered normative, as standard templates that we can lay on a real object as a way of highlighting the elements needed to solve problems. Tradition, in other words, plays a huge role. Many separations among disciplines have appeared as a result of centuries-old practices. These divisions become conventional and canonical; in logical analysis, it is difficult to separate them from the objects of study, to understand that they are only tools. For example, centuries of medical practice have taught us to view the human body as a system of organs: brain, kidneys, liver, etc. We internalize this framework as children, and it is very difficult to us to recognize that if medical practice were different the body's division into elements would be different, i.e., the human body might consist of wholly different organs.

New problems require new kinds of analysis. We should not consider non-traditional system representations "unrealistic" until they can be affiliated to the traditional ones; they are just different, because the problems they address are different.

One of the problems connected with the choice of an analytical framework is a question which, on the face of it, seems out of field: what is a human being? The answer may be obvious from a biological or spatial-temporal viewpoint. If, however, we consider the human being as a functional element in a social organism, it appears only through its external connections with technology, mode of life, semiotic tools and, finally, activity. Thus another mode of analysis is necessary, not spatial-temporal but functional, which would allow us to represent the human being as an element of a social organism.

Perhaps, we need to try out our analysis on system entities closed with respect to certain activities, so that the entities would contain these activities as internal functions (Pospelov, 1969). The human organism initially appears to be the operator with respect to such entities, and activity appears to be human manipulation by means of semantic and of technical tools.

One of the tendencies in the development of social organisms is to reduce the importance of human control. This is especially clear with respect to machine production, where human control has largely been replaced by automation.

It is necessary to emphasize that machines' inner functioning (in particular that of electronic machinery) is based on human activity only in historical terms. After a certain point, they begin to evolve according to their own laws and to acquire properties and structures completely alien to the properties and structure of the initial activity on which they were patterned. In order to describe this functioning in terms of the social organism and to predict the fate of its biological component, we need a completely new set of analytical concepts.

## Configurator

Some problems require more than one model or representational framework, that is, they need more than one analysis of the whole into its parts. Sometimes the problem can be solved only by using several different but interconnected system representations producing different functional elements, as if the object were being projected onto several screens. Each screen gives its own articulation into elements, thereby generating a different structure for the object. The screens are connected in such a way that the researcher can correlate different parts of the representations. A device that synthesizes different system representations will be called a *configurator* (Lefebvre, 1962).

A configurator is a special structure that can be found in all areas of knowledge. For example, in radio technology there are several ways of representing a system: as a flowchart, a circuit diagram, or an assembly diagram of the same device. The flowchart reflects the technological units constituting the system. The circuit diagram is different. It must explain the functioning of the device. Devices may have different flowcharts but the same circuit diagram. Finally, the assembly diagram represents the structure of the device with a view to its construction. Only the synthesis of these representations gives us all necessary knowledge about a subject. It is impossible to say what elements a given device consists of without referring to the system representation. The representation of an object with a configurator is widely used in descriptive geometry. Any sketch of a component in three projections (plan, face, side view) represents the result of using the Cartesian system of coordinates, which is also a kind of geometrical configurator.

## System Representations and the Object in a Researcher's Reflexion

A scientist may recognize system representation as repetitive patterns. To do this, the researcher needs a way to register his own internal system representations. In addition, the researcher needs a special tool for depicting the object (sf. Shchedrovitsky, 1966). The depiction enables us to compare different pictures of the object. In order to say, "these are different pictures of the same object," we must have a way of defining "the same." The system representations and the object in the researcher's reflexion are given in Fig. 52.

The left-hand rectangles are different depictions of what are, generally speaking, different objects. They become different depictions of the same object in the researcher's reflexion when connected with the reflexive picture in the little man's left hand. The "itinerary" of the researcher's thought process, moving from one position to the other, is also depicted in Fig. 52. It is our symbolic depiction of relationships between the positions. The properties and qualities of the objects obtained by the researcher fall into two classes in his internal world.

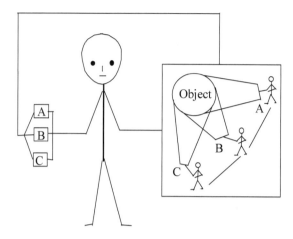

Fig. 52

Some of them relate to the object and are "attributive," i.e., they characterize the object *per se*; the others, from the researcher's point of view, are generated by patterns A, B, and C, the patterns through which he looks at the object and which determine different positions. We designate them as $x_1$, $x_2$, $x_3$ respectively.

If the external position of the little man is Y, then the entire situation corresponds to the following reflexive polynomial:

$$(T + Tx_1 + Tx_2 + Tx_3)y \ .$$

Thus, the configurator can be depicted as a reflexive polynomial. Consider a few situations. Let $Tx_1$ be the schematization of a situation by cybernetics, and $Tx_2$ by physics. There are four possible cases:

$$\Omega_1 = T + Tx_1 + Tx_2 \ ,$$
$$\Omega_2 = (T + Tx_1 + Tx_2)\, x_2 \ ,$$
$$\Omega_3 = (T + Tx_1 + Tx_2)\, x_1 \ ,$$
$$\Omega_4 = (T + Tx_1 + Tx_2)\, x_3 \ .$$

In the first case, the researcher does not have a holistic picture; he is not aware of the means at his disposal. The world appears to him in two ways: on the one hand, as a huge cybernetic machine, and, on the other, as reality subject to the laws of physics. The researcher makes no connection between the cybernetic machine and the physical reality.

In the second case, the entire situation becomes conscious from the physical point of view, i.e., a picture generated by cybernetic patterns of analysis is reduced to physical models.

The third case reflects the cybernetic point of view; this means that $Tx_2$ is reduced to $Tx_1$.

The fourth case - creation of a new point of view - is typical for scientific work. If we use an analogy to reflexive games, this process is the construction of a new player who can become aware of pictures constructed by other players[1].

We may suppose that scientific knowledge can be schematized in the form of a reflexive polynomial, whose terms correspond to different points of view. In this sense, connecting to a "body of science" is the foundation for studying the reflexive object. Education would involve trying out different points of view, and creative activity would be an aggressive act involving the entire structure: eliminating some terms, introducing new ones, constructing families of cooperating and competing viewpoints.

The role of "the object as such" is played by $T$ inside the parentheses. Yet from the external investigator's position $T$ is also a specific system representation of an object, for example, from $X_3$'s position. This system representation possesses the "objective privilege," which the other system positions made conscious of through the person of the investigator as deduced from $T$.

For example, astronomers use two completely different paradigms: representing the Sun and the planets as a heliocentric system, and at the same time representing them as objects attached to the celestial sphere. The mechanical analogues of these theoretical paradigms are the tellurium and planetarium. One of them (tellurium) is considered true, while the other (planetarium) is considered reducible to the first and used only for convenience. If the planetarium position is designated as $x_1$ and the tellurium position as

---

In this way we can depict a methodological investigation. Designate the methodologist's position as $x_1$, so the situation is $(\Omega_1 + \Omega_2 + \Omega_3 + \Omega_4)x_1$. The methodologist is aware of the logical mechanism of reducing one system representation to another.

$x_2$, the connection between them is $(T+Tx_\mathrm{i})x_2$.

In recent decades, physics has begun playing the tellurium-role. Many biologists, cybernetics, and chemists are convinced that genuine and exhaustive knowledge in their respective areas can be achieved only if they succeed in reducing all laws to physical ones. This situation can be explained on historical grounds. Physics was already highly developed at a time when, for example, biology did not even exist as a unified science, and the concept of complex system shall not yet arise. When physicists were constructing cosmological models, they cared little that these models would later have to be colonized with biological objects, with mind and civilizations, perhaps even with more complex objects. The models constructed by physics are not easily adaptable to such colonization. And the reason is not the well-known second law of thermodynamics, which is considered to be the main enemy of biological objects, but rather in the specifics of physical models. They are simply not designed for the analysis of biological objects, mind and civilizations.

## Organism as a Gas, Organism as a Technical Device

The first approaches to investigating biological objects were based on their representation as a peculiar "ordered gas." This allowed researchers to analyze them using concepts of thermodynamics and the molecular-kinetic theory of gases. In such approaches, the degree of organization (организованность) is reduced to the quantity of entropy in the system.

Another approach to investigating biological objects originated in engineering, with its extensive system of concepts, schematic diagrams, wiring diagrams, flowcharts, and functional diagrams of various kinds. Special types of calculus were devised as well as methods of translating from one type of depiction into another, i.e., special configurators were constructed.

In 1940s, a fundamental change occurred in the methods used for studying complex objects: the means that were previously used to design machinery began to be used for an entirely new purpose: to model real objects. The engineering paradigm became a tool for scientific thinking, possibly because, during the Second World War, researchers from the sciences - physics, chemistry, biology - were drafted to work in military engineering. They arrived having scientific skills, but were forced to acquire a knowledge of technical design. After the war they found themselves in possession of two different tool sets with which to face old scientific problems.

These engineering skills proved extremely useful, for two reasons:

1. The mechanisms they could model were extremely diversified. For this reason, the set of semantic elements from which depictions were composed also had to be diverse and to allow for a huge number of variations in how the elements were combined. This made it possible to depict objects which previously could not be represented as systems due to lack of the necessary visual tools.

2. Many devices needed to be represented as several different systems, for example functional diagrams showing the operational cycle a given system, or wiring diagrams showing the spatial relations of the elements. This allowed designers to select different angles and depths of modeling and to synthesize the views into a single whole.

Treating an organism as an ordered gas had practically become obsolete, because it does not capture structural or functional aspects. Yet this approach reappeared in research into the nature of organization even without respect to special systems, i.e., when general concepts of organization and self-organization are being considered. In this type of study, the concept of organization (организованность) is taken as intuitively obvious, and the concept of entropy is developed in such a way as to substantiate the intuition. Therefore, we observe a gap between "theoretical practice," using the tools of engineering, and "theoretical awareness," in which thermodynamic concepts not at all connected with engineering are also used. Because of this, there is no general notion of organization adequate to the structural-functional features of complex organisms.

## Principle of Borrowing

What is usually meant by organization? What is the intuitive understanding of organization and order? In his paper "On self-organizing systems...," Foerster provides an example illustrating one point of view on self-organization (Foerster, 1960). A set of magnetized cubes is placed in a box and the box is shaken. The cubes, which had originally lain in disorder, become organized into a nice geometrical composition. Foerster considers it obvious that the organization of the system was higher after shaking. Indeed, the cubes do seem more organized after being shaken, yet our intuition resists this conclusion. The problem is that we are using canonical standards of order. We compare an object with the standard, and if they coincide we say that the object is organized or ordered. Foerster calls his arrangement an extremely ordered structure that might be shown at an exhibition of surrealist art.

Here is a contrary example. An ancient pattern looks like a chaotic conglomeration of lines and dots to the uninitiated, but an archeologist might compare it with a sample of ancient writing and discover the pattern of a text. Our intuition is based on our standards, and we do not control which standards are used. Moreover, as a rule, we do not distinguish our standards from the results of comparing them with real objects. The same object may appear organized by some standards and a disorderly heap by others. A process, examined by certain standards, increases organization, while by other standards it decreases organization. Relying only on intuition, we will never be able to identify self-organizing systems in such a way that the process of self-organization becomes their true attribute; we can never be sure that our research tools were only the means of detecting this attribute and not the means of imposing it. If we want to avoid misconceptions, we need to devise a special procedure to identify the characteristic of self-organization.

For example, we wish to find at least one objective property in the triangle $ABC$. The length of sides $AB$, $AC$, or $BC$ cannot be such a property, because it is determined by our choice of the unit of length, which is arbitrary. If we choose one of the triangle's sides as a unit, i.e., we borrow one of the triangle's elements as our tool and measure the other sides with its help, we will obtain an objective characterization. In fact, by considering the ratio of the measures of two elements of the whole we always move to attributive properties, because this procedure removes the results of using an arbitrary standard of measure and leaves only a pure ratios of parts.

We may construct a similar procedure for identifying the parameter "degree of organization." This parameter characterizes only systems whose structure permits borrowing a standard of organization from them. The principle of extracting the means for our evaluation of the system's degree of organization from the system itself will be called *principle of borrowing* (Lefebvre et al., 1965).

We will call objects *organizing* if they can be presented as two elements plus the mechanism of a link signifying that the structure of element $A$ is transferred to element $B$, i.e., $A$ is a sample or a design for structuring $B$.

After identifying an organizing system we can claim the attributive property of organization in relation to element $B$ (*but not to the whole system!*). For this we must use the principle of borrowing: we must extract, from the system, the structure of element $A$, which is used by the system as a sample and, after reconstruction, use this structure for our own purposes as a sample of organization. This standard can be applied to $B$.

The measure of $B$'s deviation from the sample will be called $B$'s *dissonance*. For example, dissonance in a student's dictation assignment is the deviation of the student's text from the authoritative text used by the teacher. We have to borrow a sample from the system and use it as our means for measuring dissonance. A teacher usually performs two functions. During the dictation, the teacher and the student constitute an organizing system. When the teacher is evaluating the student's work, the teacher is an investigator, who borrows an element of the system, which he was earlier part of, as his investigative tool. By this reasoning we can characterize the organization of only one element of the system.

To determine the organization of the whole, we have to introduce the concept of a *self-organizing system*. We will call objects self-organizing if they can be represented in such a way that one of their elements has the function of a design for the whole, i.e., this element contains a certain structure. A special mechanism produces the structure of the whole following the sample of the structure of this element.

The principle of borrowing allows us to understand organization not only as the characteristic of an element in a system, but also as a characteristic of the system as a whole. In borrowing a design as a means for the system representation of the whole, we can evaluate its organization by constructing a special measure of dissonance, a measure of the whole system's deviation from its own internal design.

This approach allows us to introduce the concepts of organization and self-organization, avoiding an absolute "worldwide" standard for the organization, whose function is often given to the measure of entropy.

## Conflict of Structures

To adequately describe complex real systems (social or biological), representing them as organized systems with one design is not enough. Consider, for example, a chess game. Each partner, *A* and *B*, has his own inner tablet where he maps positions on the board and plans his actions. Obviously, an arrangement of chess pieces that is advantageous for one partner is disadvantageous for the other. The task of each player is to develop the structure in the direction of his own ideal, i.e., toward a situation in which the opponent's king is checkmated.

We may consider this entire system as organized by two competing designs. Two different structures, instantiated in the same material, are present simultaneously on the chess board. While one structure's dissonance is growing, the other structure's dissonance diminishes, and *vice versa*. It is impossible to characterize the organization of this system using a single parameter, since we need to measure two dissonances.

The same chess game can be considered as an element of another organized system if we supplement it with an additional element, rules of chess. By borrowing the rules as a sample of organization we can determine the dissonance of the process (deviations are moves forbidden by the rules). Depending on our position (say, as a referee or as a fan of one of the parties), we represent the system differently and borrow different samples for evaluating its organization. This example shows how organisms from diverse "dimensions" are incorporated into a single structure.

A complex organism appears for us as a special symbiosis of various structures realized in the same material. In one morphological body "live" several different functional systems. They live in fact, and do not merely appear as a result of the special viewpoint of an external observer.

We can illustrate our idea with an example that appears in many popular books on psychology. Figure 53 represents two images formed by the same lines: a human profile and a mouse. We can read this drawing twice; what we see is determined by our schematization. Imagine now that the mouse and the human profile live their own distinct lives. They (not an external observer) look at themselves, feel their integrity and, in addition, try to change the configuration of their parts. The mouse, for example, may twist its tail and so make wrinkles on the neck of the human profile. To keep their existence as a mouse and a profile, they must keep certain commitments to one another. If one of them changes and saves its own features, it may destroy its partner.

Fig. 53

How does the above construction differ from a configurator? In introducing the concept of configurator, we move the researcher to the outside. Now we have constructed an abstract object in which several constructors-researchers are made from the same material. The process of seeing the object is closed to the object itself. Only this allows us to introduce the attributive property of degree of organization. An ideal object that has several structures will be called a *configuroid*.

In the next Chapter we consider a model containing the idea of a configuroid.

**Chapter VIII**

# Janus-Cosmology

Self-organizing systems are not part of the world view of physics. Giant cosmic civilizations are assumed to exist, but always as something outside of properly natural processes. Contemporary cosmological models are generated by physics; biological objects create discord in the resulting world view.

A new cosmology, fundamentally different from the physical one, has been taking shape over the last few decades. Its defining trait is to integrate biological reality into the world view as an essential component. The new cosmology is associated with scientists such as John von Neumann, M.L. Tsetlin, E.F. Moore, W.R. Ashby, L.L. Lofgren.

It seems appropriate to consider possible models for this cosmology and how they might be constructed. One aspect of these models would be living organisms and civilizations, while the other would be physical phenomena, as two different manifestations of a single unified construction. In constructing physical models, particularly cosmological, it is commonly assumed that order and chaos are absolute phenomena, not dependent on the observer's principle of organization or the observer's cognitive tools. We will suspend this assumption and demonstrate the possibility of constructing a model of the symbiosis of two different self-organizing systems made from one material, i.e., a configuroid with two structures. The observer's evaluation of phenomena as organized or disordered will depend on which side of the configuroid the observer belongs to.

## The Main Idea

The algorithm of a self-generating automaton is based on the spatial relations of its elements. This algorithm now performs the new function of an explanatory tool of spatial rearrangement. Unlike previous explanations of spatial rearrangement within kinematic-dynamic models, in Neumann's self-generating automaton, the spatial movements may be interpreted as results of the functioning of the algorithm itself. Usually, this aspect is not considered, and the evolution of the automaton is seen in terms of maintaining its power supply, i.e., in traditional dynamic terms.

The opposite type of reasoning is also possible. The algorithm may be taken as primary, and the dynamic picture deduced from it through a particular interpretation of the automaton's obedience to laws of logic. The author has called this automaton a *Janus-Cosmology*.

To clarify this idea, let us recall the sliding tile puzzle. There are fifteen tiles, marked 1 to 15, in a box measuring 4x4, with one position vacant. Initially the tiles are in a disordered state. The player's task is to order them. Imagine that the tiles are numbered on both sides and the numbers on either side of any tile are different. Two players located on opposite sides of the box play the game and are not aware of one another's existence. For each of them, the activity of the other player would appear as chaotic movements by the tiles generally tending to destroy orderly configurations. This construction allows us to introduce an analogue to energy. Assume that the players are not allowed to damage each other's orderings. If a local ordering on one player's side damages the situation on the other player's side, the other player must be allowed to gain order in another location.

An observer looking at one side only would note that, in order to obtain increase order in area $C$, disorder must increase in area $D$; the observer would interpret this fact as the transfer of energy from $D$ to $C$.

## Processes of Self-organization on a Moebius Strip

The kind of interaction described above is of special interest when the interactions take place on a one-sided surface such as a Moebius strip. On a one-sided surface, it is impossible to partition all systems into two disjunct classes. Some systems may be both opposites and at the same time neighbors. Let us take a piece of a tape divided into cells on both sides and number the cells (Fig. 54), then glue this tape into a Moebius strip (Fig. 55). If a system "living" on the surface of the stript wants to put the numbers in order, it faces an insoluble problem.

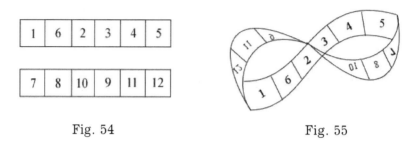

Fig. 54                                    Fig. 55

## Cellular Automata on One-Sided Surfaces

Lofgren (1958) examined the process of self-reproduction with the assumption that there is some probability of an element's transferring into a faulty state. This probability is superimposed on the element from the outside; it does not arise from the element's construction. In fact, it was assumed that the elements are statistical in nature. The models of Janus-Cosmology allow us to proceed in another way. On the one-sided surface, it is possible to design a self-reproducing and strictly determined cellular construction whose elements, from an observer's point of view (an observer who is not aware of the existence of the opposite sides), are statistical in nature.

Consider a Moebius strip divided into cells $u_0$, $u_1$, ..., $u_{2n-1}$. The cell following $u_{2n-1}$ is $u_0$. Each cell can take one of two values: 0 or 1. The state $u_i=1$ will be interpreted as the existence of an organism in this cell. To introduce relations between opposite cells, we use the operation modulo $2n$: the opposite of cell $u_i$ will be designated as $u_i^{-1}$ and considered to be equal to $u_{i+n}$.

Let us use a discrete time scale and assume that the state of each cell at moment $t$ is fully determined by its own state, the states of its neighbors, the state of its opposite and its neighbor's opposite's states at the moment $t$-1.

Suppose that the organisms may reproduce, i.e., under certain conditions, new organisms may appear in the neighboring cells. The formal rules of this "universe" are as follows: If, at moment $t$, $u_i=0$ and at least one of its neighbors is in the state 1, then at moment $t+1$ the value $u_i=1$, otherwise $u_i=0$. If, at moment $t$, $u_i=1$ and more than one antipode is in the state 1, then at moment $t+1$ $u_i=0$, otherwise $u_i=1$. Consider a system consisting of six cells. Their initial states are as follows:

| 0 | 1 | 0 |
|---|---|---|

$u_0$   $u_1$   $u_2$   $t$

| 0 | 0 | 0 |
|---|---|---|

$u_3$   $u_4$   $u_5$

In the next moment, the system expands:

| 1 | 1 | 1 |
|---|---|---|

$u_0$   $u_1$   $u_2$   $t+1$

| 0 | 0 | 0 |
|---|---|---|

$u_3$   $u_4$   $u_5$

In the next moment only one cell becomes empty:

| 1 | 1 | 1 |
|---|---|---|

$u_0$   $u_1$   $u_2$   $t+2$

| 1 | 0 | 1 |
|---|---|---|

$u_3$   $u_4$   $u_5$

Then annihilation occurs; all the organisms disappear, except for one:

| 0 | 0 | 0 |
|---|---|---|

$u_0$       $u_1$       $u_2$       $t$

| 0 | 1 | 0 |
|---|---|---|

$u_3$       $u_4$       $u_5$

Furthermore, our "universe" will perform an exactly analogous evolution in the opposite direction: it will expand, then be annihilated, then 1 will appear in cell $u_i$ and the cycle will begin again.

Imagine a scientist in this "universe" who is not aware of the existence of "opposites." This scientist believes that a cell's state at moment $t$ is a function of this cell's state and its neighbors' states at moment $t$-1. As Moor (1964) points out while considering an ordinary cellular automaton, this corresponds to the assumption that the interaction cannot take place at a speed greater than the speed of light. Our scientist cannot use a deterministic model; he will realize that the preceding states of the cell's neighbors do not always determine the cell's new state unambiguously. For example, when the neighbors of a 'one' are 'ones', in one case out of four the 'one' becomes 'one', but in the other three cases the 'one' becomes 'zero'. The researcher is forced to introduce a law of distribution: with the given neighbors, the cell behaves according to expectation only on average: it transfers to 'one' with probability 1/4 and to 'zero' with probability 3/4.

The development of cellular structures on one-sided surfaces has its own particular mathematical interest. It is easy to sketch a two-dimensional cellular construction. Consider the following square divided into cells:

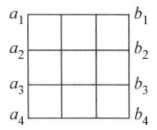

Suppose that its reverse side is also divided into similar cells and that it is possible to deform this square in any way and make a hole in the central cell without disrupting its operations. Glue line $a_1b_1$ to line $a_4b_4$ to obtain a cylindrical surface, then turn a part of this cylindrical surface inside out and pierce it through the hole, and glue the edges (circles) of the cylinder so that line $a_1,a_2,a_3,a_4$ will coincide with line $b_1,b_2,b_3,b_4$. This gives us a Klein bottle (Fig. 56).

Fig. 56

Each cell has eight cells that are its neighbors and frame it. Let us introduce rules of reproduction and annihilation similar to those for the one-sided case. If a cell is in state 0 and at least one of its neighbors is in state 1, then at the next moment 1 will appear in that cell; otherwise, the cell remains in state 0. If a cell is in state 1 and more than five opposite cells are in state 1, then at the next moment the given cell acquires 0; otherwise, it remains in state 1.

We will depict the given structure as two tables; cells in the same locations are opposites. The "civilization" on this surface will evolve as follows:

**$t$**

| 0 | 0 | 0 |
|---|---|---|
| 0 | 1 | 0 |
| 0 | 0 | 0 |

| 0 | 0 | 0 |
|---|---|---|
| 0 | 0 | 0 |
| 0 | 0 | 0 |

**$t+1$**

| 1 | 1 | 1 |
|---|---|---|
| 1 | 1 | 1 |
| 1 | 1 | 1 |

| 0 | 0 | 0 |
|---|---|---|
| 0 | 0 | 0 |
| 0 | 0 | 0 |

**$t+2$**

| 1 | 1 | 1 |
|---|---|---|
| 1 | 1 | 1 |
| 1 | 1 | 1 |

| 1 | 0 | 1 |
|---|---|---|
| 1 | 0 | 1 |
| 1 | 0 | 1 |

**$t+3$**

| 0 | 0 | 0 |
|---|---|---|
| 0 | 0 | 0 |
| 0 | 0 | 0 |

| 0 | 1 | 0 |
|---|---|---|
| 0 | 1 | 0 |
| 0 | 1 | 0 |

**$t+4$**

| 0 | 0 | 0 |
|---|---|---|
| 0 | 1 | 0 |
| 0 | 0 | 0 |

| 1 | 1 | 1 |
|---|---|---|
| 1 | 1 | 1 |
| 1 | 1 | 1 |

**$t+5$**

| 1 | 0 | 1 |
|---|---|---|
| 1 | 0 | 1 |
| 1 | 0 | 1 |

| 1 | 1 | 1 |
|---|---|---|
| 1 | 1 | 1 |
| 1 | 1 | 1 |

| 1 | 0 | 1 |
|---|---|---|
| 1 | 0 | 1 |
| 1 | 0 | 1 |

| 0 | 0 | 0 |
|---|---|---|
| 0 | 0 | 0 |
| 0 | 0 | 0 |

$t+6$

The system's evolution continues cyclically. The system's state at moment $t$ is determined by its state at moment $t-1$. The system is closed. It has no neighbors which would influence its transition from one state to the next.

## Janus-cosmology and Time Arrows

The schemes that we have considered assume that Newtonian time is common to the entire system. Cosmological models using one-sided surfaces allow construction of various alternatives. Consider one more. Suppose that the organization of the entire system is constant. This means that an increase of organization in one location is compensated by a decrease in some other location. Depict the system once more as a Moebius strip divided into cells:

| | I | | II | | III | | IV |
|---|---|---|---|---|---|---|---|
| 16 | 1 | 2 | 3 | 4 | 5 | 6 | 7 |
| | V | | VI | | VII | | VIII |
| 8 | 9 | 10 | 11 | 12 | 13 | 14 | 15 |

The lower strip is the reverse side of the upper strip. The total organization of any two cells-opposites is constant; under this condition, the entire system's degree of organization is also constant.

Let an observer sit in each cell that contains a Roman numeral; the observer registers the state of his own cell and the states of the two neighboring cells. Suppose that each observer has discovered that in his "world" the second law of thermodynamics holds, i.e., that organization in his cell decreases. This statement corresponds to the assumption that, in any large area, entropy increases. This leads us to conclude that in opposite cells time flows in opposite directions. Indeed, if the organization registered by the observer in the cell I decreases, then, since the total organization of opposing cells is constant, the organization registered by the observer in the cell V must increase, which means that the time there flows backwards.

On the other hand, if the observer in cell I can communicate with the observer in cell II across the empty cell between them, they will conclude that their time flows in the same direction. If observer II communicates with observer III, they will also conclude that time flows in a single direction. It turns out that, for each pair of neighbors, time flows to the same direction. By communicating in this way the observers would conclude that everywhere time flows in the same direction. Nonetheless, we see that there are pairs for whom time flows in different directions.

This paradox can be resolved by assuming that the rate of time slows with distance, stops, and begins flowing again in the opposite direction.

Designate a time interval as $\Delta t$ and consider the following function:

$$\Delta t = \frac{1}{\cos(\frac{\pi}{n})k},$$

where $n$ is half of the even number of cells $N = 2n$, $k$ is the distance from a given cell to another. The cell is located at $k = 0$ from itself, at $k = 2$ from its neighbor, etc. We have sixteen cells, thus

$$\Delta t = \frac{1}{\cos(\frac{\pi}{8})k}.$$

Begin with the first cell. By setting $k = 0$ for that cell, we obtain $\Delta t = 1$. For $k = 2$, $\Delta t = \sqrt{2} > 1$; for $k = 4$,

$$\Delta t = \frac{1}{\cos\dfrac{\pi}{2}} = \infty ,$$

i.e., in the fifth cell, where observer III is located, time stops in relation to cell 1 (but only in relation to the cell 1). Finally, for $k = 8$

$$\Delta t = \frac{1}{\cos\pi} = -1 .$$

Therefore, in cell 9, where the opposite of observer I is located, we obtain an opposite directionality of time equal in rate to cell 1. So, each cell has its "horizon," i.e., a cell where time stops in relation to the given cell and begins to flow in the opposite direction.

Let us continue our speculation. In the "universe" we have constructed a redshift must appear, due not to the recession of galaxies but to the deceleration of time. Light from objects that are beyond the horizon cannot reach the observer, because, from his point of view, it goes in the opposite direction, back to its own source.

This model has some resemblance to de Sitter's model, except that the spatial frame is taken to be a one-sided surface rather than a sphere. This allows us to identify spatially distant places and associate a "warping" of time with degrees of organization.

# Chapter IX

# Systems Drawn on Systems

The central task in studying complex objects consists in elaborating models of reality in which ideal and material phenomenologies are interconnected. The possibility of considering systems endowed with intellect as unified systems depends on the solution to this problem; otherwise, we will have to be satisfied with two separate directions of study. Apparently, we need to construct a special configurator whose various particular schematizations must become projections of an ideal object and correspond to different phenomenologies. New concepts are needed to work on this task.

In this chapter we will try to outline a series of techniques which, in our opinion, may be useful in constructing the configurator in question.

## Organism and Substance

In speaking of systems we often assume that there is some substance that constitutes them and gives them life. The first difficulty with this appears in attempting to describe a simplest living organism. Wiener once said that the individuality of the body is more like the individuality of a fire than of a rock; it is the individuality of a structure rather than of a piece of matter. An organism as a whole is not reducible to atoms. We are dealing here with a reality that is holistic, no less real than a rock, but not understandable in terms of static material consistency.

## Organism as Wave

There have been attempts to construct functional models of living organisms by representing them as cellular automata. Each element of such a structure is able to be in a finite number of states. The configuration of the active states of the cells describes the organism. It is possible to construct a moving organism that would spread like a wave (Lofgren, 1958). In the previous chapter we examined one version of such a model. These automata reproduce some features of living systems and permit explanation of some of their processes. Although such models succeed in breaking the dependence on atoms, we nonetheless have to deal with the substance in which an organism moves as a wave. The substance remains primary, the wave secondary. If there is no substance, there is no wave.

## The Relation Tissue-pattern

In a short story (Platonov, 1964), several thousand people were made to line up on a playing field, where each of them functioned as a component of a digital computer (there is nothing inherently impossible in this situation). Imagine now that this computer, constructed from human organisms, functions for a number of years. During this time, the matter in these organisms changes; they consist of other atoms but remain the same people (as if organisms move along substance). Imagine now that the people change their functions every day; for example, today everyone performs the function which his neighbor was performing yesterday. It appears that this computer is moving through the field of human organisms, yet with these changes in substance nothing changes in the computer's structure. Thus, we have constructed a persistent functional structure, which moves through a sustained functional structure, which in its turn, moves through a substance of atoms. The functional network of human organisms serves as the computer's substratum, and the set of atoms serves as substratum for the human organisms.

The human bodies relate differently to the sets of atoms and to the computer. Each body is a functional system in relation to the sets of atoms, and it is a dead substratum in relation to the computer.

The relation between a substratum and a functional schema will be called *tissue to pattern*. It is as though a functional schema were drawn on the substratum. It is not similar to a pattern on a rug, but rather like a moving picture on a screen. Another picture may be drawn on top of the first picture. For example, a movie frame may contain a picture of a movie theater in which a film is being shown. A movie in a movie is a pattern on a pattern: the tissue is the movie on the screen.

The relation of tissue to pattern was used by Stanislaw Lem (1964) in a fictional story where a constructor was constructing civilizations, in which new constructors appear and were constructing civilizations with a constructor in each, and so, on until there appeared a civilization in which all the inhabitants were happy.

## Closed Chains of Tissue-pattern Relations

Relations of tissue to pattern may form chains (Fig. 57). The arrows indicate the relation "tissue-pattern." The elements in this structure are not equal. For example, element 1 (symbolizing atoms) functions only as tissue. It is matter itself, not drawn on any other deeper substratum. After completing the scheme in Figure 57, the author feels an irresistible desire to deprive element 1 of this privilege. To do this, the author closes the chain, i.e., determines element 1 to be drawn on element 4 (Fig. 58). Now all the elements are equal; each of them performs two roles.

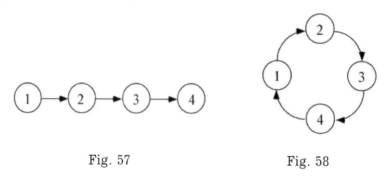

Fig. 57                                    Fig. 58

A new question now arises: is this closure merely a formal notion? Can we conceive of a meaningful construction that would function with such a circular structure?

## Closed Chains of Automata Drawn on Each Other

Let a field of cells be given (Fig. 59). Each cell can be in one of four states: $a_1$, $b_1$, $c_1$, $d_1$. The structure of four elements will be considered an organism that occupies the entire field (Fig. 60). Suppose that each state is a self-reproducing system: $a_1$ is produced in the cell where the preceding state was $b_1$, which is produced in the cell of $c_1$, which appears in the cell of $d_1$, and, finally, $d_1$ replaces $a_1$. It is easy to see that if the system is left to itself, it will rotate in a clockwise direction.

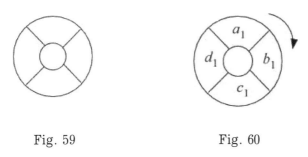

Fig. 59                    Fig. 60

The cells, each of which is always in exactly one of the four states, represent the tissue, and the rotating automaton is the pattern.

Let us allow each of the states $a_1$, $b_1$, $c_1$, $d_1$ to be in four states $a_2$, $b_2$, $c_2$, $d_2$. Construct a new automaton and set it on top of the rotating automaton. The rules for state-production remain the same. System $a_2$, $b_2$, $c_2$, $d_2$ begins rotating over the automaton $a_1$, $b_1$, $c_1$, $d_1$, which plays the role of the tissue in relation to the former. In relation to the paper where the cellular area belongs, the automaton $a_2$, $b_2$, $c_2$, $d_2$ will jump over one cell in each change of state. Indeed, the automaton $a_1$, $b_1$, $c_1$, $d_1$ rotates one cell in relation to the paper, and the automaton $a_2$, $b_2$, $c_2$, $d_2$ rotates one cell in relation to $a_1$, $b_1$, $c_1$, $d_1$. Now set automaton $a_3$, $b_3$, $c_3$, $d_3$ on top of $a_2$, $b_2$, $c_2$, $d_2$ and automaton $a_4$, $b_4$, $c_4$, $d_4$ on top of $a_3$, $b_3$, $c_3$, $d_3$.

In relation to the paper, automaton $a_3$, $b_3$, $c_3$, $d_3$ will jump two cells and the automaton $a_4$, $b_4$, $c_4$, $d_4$ three, that is, it will be stationary. For the observer who views the automaton as immobile, the automaton $a_3$, $b_3$, $c_3$, $d_3$ will rotate in the opposite direction. This effect is analogous to the effect of car wheels rotating backward in movies.

The example with the automata drawn on top of one another demonstrates the relativity of the concept "material." An element that is material in relation to one element may be drawn in relation to another.

## Pattern as Sustainable Change

The terms "tissue" and "pattern" raise questions: how is the pattern drawn on the tissue? What is the nature of the coloring? We will consider the "paint" or "coloring" to be a consistent pattern characterizing the tissue.

We assume that the principal distinction between living organisms possessing intellect and machines is that organisms have long and sometimes closed chains of tissue-patterns and that each pattern is a self-organizing system[1]. An ordinary machine is a chain consisting of one "tissue-pattern." What we call the Physiological is a pattern drawn on "the set of atoms," and what we call Mental is a pattern drawn on the physiological.

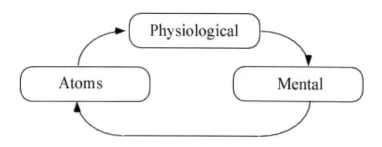

Fig. 61

---

[1]In fact, in the previous chapter we already used the relation tissue-pattern in constructing the Janus- Cosmology. We mapped two patterns on opposite sides of the same piece of paper. After we glued the Moebius strip, we turned them into a pattern contradicting itself.

Further elaboration of this model may proceed in one of two ways. We may construct hierarchies of systems with the relation tissue-pattern, or we may produce a circuit. Then, the sets of atoms represent a pattern drawn on the psychological and, at the same time, a tissue for the mental (Fig. 61). In a unified ontological picture, mind and body are interconnected.

The relation tissue-pattern may be useful in the study of complex processes. Presently, there is an abyss separating the phenomenology of mental processes from the phenomenology of physical processes; existing concepts are powerless to bridge it.

The relation tissue-pattern is logically opposed to the traditional relation "part-whole," dominant in most theories of complex systems. It was Democritus who first conceived of absolutely substantial element-atoms of which bodies and events were composed. Their properties were derived from atoms, which already contained the qualitative attributes of the self. In the tissue-pattern construction, the pattern is not a part of the tissue; it is a phenomenon developing according to its own laws. The connection between tissue and pattern reminds us of the connection between the text of moving advertisements ads and the set of electric bulbs on which these ads run. No matter how well we may understand the electric circuits controlling the display of the text, it does not help us to understand the text itself, its logical and linguistic structure and its meaning. Yet without the electric lights the text does not exist, and smashing the bulbs will necessarily extinguish the text.

Attempts to resolve the psychophysical dualism by reducing mental to physiological phenomena are based on a firm belief in atoms as the sole self-sufficient reality. All attempts of this kind have failed; their reductionism inevitably elides the phenomenon of mind. Thus we must seek new models of reality and construct new ways of reasoning.

# Concluding Remarks

Imagine a Box that communicates with the researcher. The Box is a special object-partner. In communicating with the Box, the researcher enters into a relationship of parity with him. The researcher is interested in such questions as, Does the Box understand him? How and to what extent can the Box predict the researcher's thought? How can the researcher receive the information he is interested in? The Box might simply answer the researcher's questions. If the Box does not speak, the researcher might try to interpret the Box's behavior, i.e., switch to a reflexive description by attributing motives, goals, and an ability to conduct substantive arguments to the Box. The researcher may have a physiological description of the Box's body, even a highly detailed one, but then he will need to interpret it, i.e., to use reflexive descriptive means.

Suppose that our purpose is to investigate the process of dreaming. We may give as detailed as possible a description of the brain's functioning during the dream, but this description would not show us the plot and drama of the dream. In order to describe the dramatic structure of the dream, we have to use literary means, which are by nature reflexive. We might try to establish a correspondence between the physiological description and the reflexive representation; if we succeeded, we would obtain a configurator with two projections of the process under investigation.

What if the Box investigates the researcher? The traditional researcher-object distinction is then no longer taken for granted. Both sides are investigators and both sides are objects; both need reflexive descriptive means. Reflexive analysis was designed for just such situations.

Long ago, researchers were divided into Physiologists and Artists. The former were armed with functional means and the latter with reflexive. The Artists were not interested in the location of internal organs, and the Physiologists were not interested in the interrelations between people. The Artists' pictures primarily registered the reflexive features of the objects portrayed, in such way that the viewer (reader, addressee) finds himself in a special semiotic space enabling pseudo-communication. Here is an example that clarifies the concept. We look at a painting where two men are pictured. The men are talking. One is suffering, and the other is gloating; we can see this from their faces. The viewer cannot hear the conversation (in a movie he might), but here he is a participant in the conversation. He is part of the semiotic environment. The communication links pass through the viewer (Fig. 62). We call these links *pseudo-communications*[1].

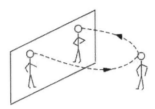

Fig. 62

The Artist's portrayal is such that the viewer may use a modification of the principle of borrowing in order to see reality through the eyes of the people in the picture. The viewer can imitate their thoughts about each other and even about the viewer, if their eyes are turned to the viewer.

The technological revolution has produced a new type of scientist armed with engineering tools. In the investigation of complex systems, scientists construct special functional pictures of the systems unlike the wiring diagrams formerly used by physicists. The models created by physics do not apply to complex biological objects and especially not to objects possessing intellect.

Then, what kind of model plays the role of a world view? In our opinion, it must be a configurator with three projections: wiring

---

[1]It is necessary to distinguish these links from the communicative connection between an artist and a viewer taking place through the picture. In fact, this connection consists in transmitting the semiotic space containing pseudo-communication.

diagram, functional schema, and reflexive. The Janus-Cosmology described in Chapter VIII allows us to demonstrate the process of constructing the configurator. We will begin with considering a construction of world view based on the Janus-Cosmology schema that would occur to a researcher equipped with the "physics ideology."

Let us take a flat sheet, make several holes in it, and insert a rod into each hole (Fig. 63). We will choose the lengths of the rods in such way that if we align their lower ends, the upper ends will stick out with lengths evenly distributed between 0 and 1. We interpret this as a low level of organization. If we move the rods so that their ends are all projected onto a small part of the interval [0,1], it is interpreted as a greater degree of organization, since the lengths of the rods' ends are distributed very irregularly on the interval [0,1]. We can similarly characterize not only the entire system but also the local areas by examining the ends' lengths' distribution in a given area. It is obvious that a living organism is an object with a particularly high degree of organization.

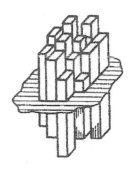

Fig. 63

If we roll up this sheet into a Moebius strip (to set a rule for the rods' relations as "neighbors"), it will become possible to project various cosmologies. For example, we may use the schema described in Chapter VIII. We can make the organization of the entire system approach the maximum and at the same time let some highly organized local systems have their own tendencies. We will probably obtain certain constructions that are interesting from the point of view of physics or biology. This cosmology, however, will never produce an intelligent researcher with ideas concerning the mechanism that generated him. Moreover, we will find only those features of living organisms that are linked to the spatial localization of matter. This is a barrier that physical models will never be able to overcome.

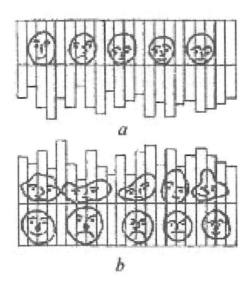

Fig. 64

Now let us construct the configurator. For simplicity, instead of rods we will use stripes that can be moved independently of one another. We shift them to form a rectangle at the top and draw little faces in the rectangle, some of them smiling, other sad (Fig. 64a). Then we move the stripes again to form a rectangle at the bottom and draw little faces in that space (Fig. 64b). By doing this, of course, we destroy the faces drawn on the top rectangle (they will acquire dissonances!). Let us initiate the game described in Chapter VIII. Following certain rules, each face can diminish its dissonance by interacting with its antipodes and neighbors. The driving force for the entire construction is the faces' desire for structural perfection.

We have supplemented the physical system of rods with the functional structure of faces. While the physical model sets conditional cosmology as a system of assembly units, we introduce functional units and set up a mechanism of connections between them and the assembly units.

Consider a researcher who looks at the functioning device depicted in Figure 64b. The researcher does not pay attention to the faces; rather, he seeks to describe the system's behavior by studying the lengths of stripes and the disposition of their upper ends. By constructing some powerful mathematical apparatus he might capture the main consistent patterns of the system. We call this researcher a Physicist. Another researcher does not look at the stripes, but rather takes the little faces as a distinct phenomenon and studies their evolution in time, correlations in structure distortions, etc. We call him a Biologist. The third researcher is interested in the facial expressions for their own sake; he identifies with each of them, tries to reconstruct their relations to one another and to reality, and constructs a typology of their expressions. We call him a Psychologist. Each of the researchers has his own object of study, his own means of representation, and his own language of communication.

Imagine now that there is a fourth researcher, equipped with a holistic configuration. He is neither Physicist nor Biologist nor Psychologist. His model allows him to describe the tasks facing the Physicist, the Biologist, and the Psychologist, because they all use projections of the fourth researcher's construction. The assembly and functional tasks are no longer separate, because the fourth researcher sees the reasons for biological functioning in the physical framework and explains the stripes' physical movements in terms of reconstructive efforts. Finally, this model has some features in common with the artist's picture. The little faces smile, the little faces mourn; i.e., a semiotic space exists in the model. The researcher may participate in a pseudo-communication link. In this world view, there are not only biological objects, but also objects comparable to the researcher in their degree of perfection. In this kind of model, the researcher eventually will be able to perceive and grasp the self.

# APPENDIX

# Iconic Calculus

The successes of applied mathematics in psychology have been modest. While the use of mathematics in celestial mechanics was a triumph both for mathematics and for astronomy, its use in the study of mental phenomena has not generated any significant psychological or mathematical ideas. The psychologist is inclined to blame the mathematician for his lack of sophistication, and the mathematician condescends to the psychologist, believing (often not without reason) that psychologists have no concept of theoretical work[1].

The mathematician is sure that his tool, well-proven in battles with Nature, cannot fail in psychology if used properly. It seems to me that the mathematician is wrong: contemporary mathematics is poorly adapted for use in psychology. It gives no measure of the content of the human mental world. So far, the human mental world has been best represented in literature and art. The language of art can have only limited use in scientific work, however, because it lacks consistency of method. In modern European culture, artistic creativity has been focused on destroying conventionality and stereotypes. While the external structure of the human mental world can be represented by some mathematical constructs, for example, reflexive polynomials, the content that makes this structure "live" cannot be shown. A special semiotic system is needed to allow representation of the human mental world in a direct way.

Below we describe a possible approach to solving this problem within the framework of the reflexive processes.

An algebraic language allows us to describe the states of reflexive systems. It represents a system's structure and the rules by which the system changes. Note that symbol $Tx$ is featureless, which

---

[1]That was written in 1973.

constitutes both its strength and its weakness. Its strength, because it is an abstract and universal representation: $Tx$ can represent the point of view of an individual, of a military headquarters, even of an entire culture. Its weakness, because it does not show anything specific. The real $X$ sees not $T$, but a reality, sometimes hostile, sometimes neutral, sometimes joyful. How can we represent this content such that it is generalized and, at the same time, conveys the full range of essential nuances?

Consider, for example, the following polynomial:

$$T + (T + Tx + Ty + Txy)z \, . \tag{1}$$

This is the detailed structural description of an individual. Symbols $T$, $x$, $y$, $z$ require interpretation. Within the framework of mathematical ideology, their meaning is not derived from their graphic representation. Algebraic symbols denote but do not depict; such is the semiotic of contemporary mathematics. Let us devise an algebraic symbol to represent the content. It must express a relation of the individual to reality. We will entrust this role to variations of the familiar "smiley" (Fig. 1).

Fig. 1

Smiley 1 expresses fear, smiley 2 - submissiveness, smiley 3 - femininity, smiley 4 - stupid credulity, smiley 5 - intellect. I am sure everybody will agree that this simple tool is an easy way to represent common emotions.

Of course, a smiley is customarily used in cartoons and has the connotation of something funny and not serious. Its repeated use in another context, however, will remove that association and make it serious enough.

We will use the smiley as a small gestalt in our calculus. It is the smallest significative graphic element. It expresses what it expresses. No verbal interpretation is necessary; sometimes such interpretation is impossible. We cannot look at a smiley indifferently, just as we cannot listen to speech in our native language as sounds without meaning. Its meaning is palpable. Smileys are similar to musical motifs in their distinctness and untranslatability to another language. With smileys, a researcher can express his relation to any subject: an individual, a social institution, nature, civilization. We will use this icon to represent the researcher's emotional contact with the object under investigation.

The idea of measuring instruments is very important in quantum physics. The situation that we are considering is even more complicated. The results relate not to instruments in general, but to the concrete "instrument" that is the researcher, "armed" with his own psyche. Psychological reality is diverse; in moving from one researcher to another, perspective changes and external reality appears differently. In moving from one culture to another, the changes may be catastrophic: some areas of psychological reality may disappear entirely.

Our main idea consists in the following: instead of the symbols $T$, $x$, $y$, $z$, we will use smileys with expressive content. Thus, polynomial (1) looks as follows:

Fig. 2

In this picture we choose arbitrary expressions which, in polynomials, will be determined by peculiarities of the researcher's perception. Reality is depicted as a rectangle; sometimes it may be omitted. To simplify notation, we write the operator of awareness $1 + x$ as only one smiley $x$. The act of awareness is shown in Figure 3. On the left side of the equation, $X$ and $Y$ are given from an external point of view, and on the right side $Y$ is an element of $X$'s mental world.

Fig. 3

With this kind of depiction we can represent operators of awareness which are impossible to represent with polynomials, as for example in Figure 4:

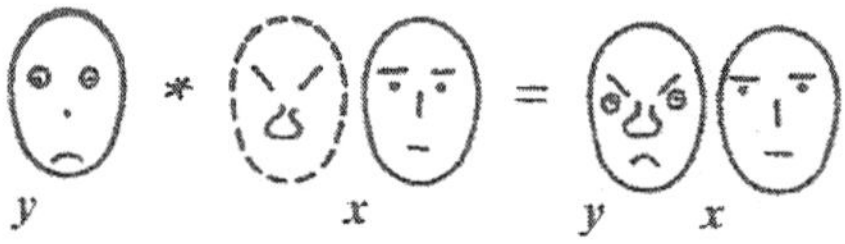

Fig. 4

The image of the other in X's mental world is predetermined; its features are shown inside a dotted oval. In Figure 4, X fills it with Y.

While using polynomials, we used only one dimension. The second dimension will allow us to register some processes of reflexive control. Suppose individual X looks as shown in Figure 5, but wants his surroundings to see him as shown in Figure 6. We depict this in Figure 7.

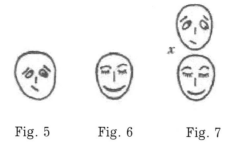

Fig. 5         Fig. 6         Fig. 7

Smiley 6 will be called a *mask*; in Figure 7 it is hanging under X's "real face".

Suppose Y performs an act of awareness. If the mask fulfills its function, it will represent X in Y's mental world (Fig. 8).

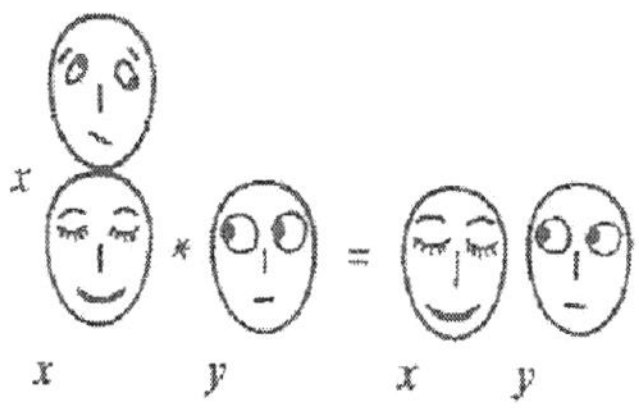

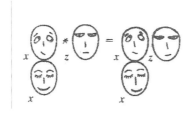

Fig. 8.

If the mask fails, the result is shown in Figure 9:

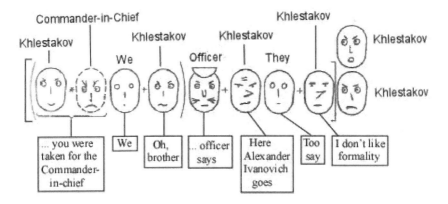

Fig. 9

To illustrate, let us use the reflexive structure of Khlestakov's monologue in Gogol's comedy, *The Inspector General* (Act III). The entire monologue is a mask (Fig. 10).

Fig. 10

A smiley is only one of many possible symbols that could be used in mathematical structures for representing emotions in a reflexive context. We might use profiles, musical notations, or abstract symbols of some kind. For the latter, a group of researchers using them would have to work out conventional explanations. Symbols that can be included in mathematical polynomials may be called *psychographics*.

The researcher of a social event must be in a relation of parity with it. Otherwise, the researcher will not understand the meaning of the elements of the situation. There are two positions for a researcher: domination and parity. Mathematical structures serve the dominating position; psychographic icons serve the parity position. To study human culture in depth, mathematics should use psychographic icons as part of its representational toolbox.

# References

Baranov, P. V. and Trudoliubov, A. F. 1969.
"On a game between a human and an automaton conducting reflexive control" (in Russian), In: *Problemy Evristiki* (Problems in Heuristic), Moscow: Vysshaya Shkola Press.

Baranov, P. V. and Trudoliubov, A. F. 1969.
"On a scheme of reflexive control independent from the plot of experimental game situation", (in Russian), In: *Problemy Evristiki* (Problems in Heuristic), Moscow: Vysshaya Shkola Press.

Blauberg, I. V., Sadovsky, V. N., Yudin, E. G. 1969.
*Sistemny Podhod,, Predposylki, Poblemy, Trudnosti* (System Approach, Background, Problems, Difficulties). Moscow: Znanie Press,

Bohr, H. 1958.
*Atomic Physics and Human Knowledge.* New York: John Wiley.

Bongard, M. M. 1967.
*Problema Uznavania* (Recognition Problem). Moscow: Nauka Press.

Chatterjee, S. and Datta, D. M. 1954.
*An Introduction to Indian Philosophy*, University of Calcutta.

Galperin, P. Ya. 1966.
"Psychology of Thinking and the Theory of the Phased Forming of Mental actions" (in Russian), In: *Issledovanie Myshlenia V Sovetskoy Psihologii* (Investigating of Thinking in Soviet Psychology), Moscow: Nauka Press.

Lefebvre, V. A. 1962.
"On Representing Objects As Systems" (in Russian). In: Symposium Proceedings "The Logic of Scientific Investigations," Kiev University Press. (English translation at algebraofconscience.org).

Lefebvre, V. A. 1965a.
"The Basic Ideas of Reflexive Game Logic"(in Russian). In: *Systems and Structures Research Problems, Conference Proceedings*, Moscow: AN USSR Press. (English translation at algebraofconscience.org).

Lefebvre, V. A. 1965b.
"On Self-Reflexive and Self-Organizing Systems" (in Russian). In: *Systems and Structures Research Problems, Conference Proceedings*, Moscow: AN USSR Press. (English translation at algebraofconscience.org).

Lefebvre, V. A. 1966.
"The Elements of Reflexive Game Logic" (in Russian). In: *Problems in Engineering Psychology*, Issue 4 (English translation at algebraofconscience.org).

Lefebvre, V. A. 1967.
"Reflexive Game Logic and Reflexive Control" (in Russian). In: *Human Decision Making*, Tbilisi: Metcniereba (English translation at algebraofconscience.org).

Lefebvre, V. A. 1969a.
"Systems Comparable to the Investigator in Their Degree of Perfection" (in Russian). In: *Systems Investigations*, Nauka Press (English translation at algebraofconscience.org).

Lefebvre, V. A. 1969b.
"Devices that Optimize Their Performance As a Result of Human Counteraction" (in Russian). *In: Problems of Heuristic*, Vysshaya Shkola Press (English translation at algebraofconscience.org).

Lefebvre, V. A. 1969c.
"Janus-Kosmologie." *Ideen des exacten Wissens*, No. 6.

Lefebvre, V. A. 1970.
"Das System in System." *Ideen des exacten Wissens*, No. 10.

Lefebvre, V. A., Schedrovitsky, Yudin, E. G. 1965.
"Artificial and 'Natural' in Semiotic Systems" (in Russian). In: *Problemy Issledovania Sistem I Struktur* (Systems and Structures Research Problems, Conference Proceedings), Moscow: AN USSR Press

Lefebvre, V. A., Baranov, P. V., Lepsky, V. E. 1969.
"Internal Currency In Reflexive Games" (in Russian). In: *Izvestia AN SSSR, Technicheskaya Kibernetika*, No. 4.

Lem, S., 1964
*Summa Technologiae*,

Leontiev, A. N. 1965.
*Problemy Razvitia Psihiki* (Problems in Psyche Development). Moscow: Mysl Press.

Lepsky, V. E. 1969.
"Studying Reflexive Processes In the Experiments Using Zero-Sum Matrix Game" (in Russian). In: *Problemy Evristiki* (Problems In Heuristic), Moscow: Vysshaya Shkola Press.

Liddell Hart, B. H. 1941.
*The Strategy of Indirect Approach*, London: Hutchinson.

Linebarger, P, M. A., 1954.
*Psychological Warfare*, Wasington: Combat Forces Press.

Lofgren, L. 1958
"Automata of High Complexity Methods of Increasing Their Reliability by Redundancy." *Information and Control*, Vol.1, No. 2, pp. 127-147.

Moor, E. F. 1964.
"Mathematics in the Biological Science." *Scientific American*, No. 9.

Platonov, K. K., 1964.
*Zanimatelnaya psichologia*, Moscow: Molodaya gvardia.

Pospelov, D. A. 1969.
"Cognition, Self-cognition and Computers" (in Russian). In: *Sistemnye Issledovania* (System Studies), Moscow: Nauka Press.

Rapoport, A. 1956.
"Some Game theoretical Aspects of Parasitism and Symbiosis," *Bulletin of Mathematical Biophysics*, Vol. 18.

Rapoport, A. 1964.
*Strategy and Conscience*. New York: Harper & Row.

Rapoport, A., Chammah, 1965.
*Prisoner's Dilemma*, Ann Arbor: University of Michigan Press.

Schelling, T. 1960.
*The Strategy of Conflict*, Harvard University Press.

Shchedrovitsky, G. P. 1960.
"Analysis of Problem Solving Processes" (in Russian). In: *Doklady APN RSFSR*, No.5.

Shchedrovitsky, G. P. 1962.
"On Distinction Between Initial Concepts in Formal and Contentual Logic" (in Russian). In: *Problemy Metodologii I Logiki Nauki*.

Shchedrovitsky, G. P. , 1964.
*Methodology of System Research*, Moscow: Znanie Press.

Schedrovitsky, G. P. 1966.
"Notes on thinking using schemes of dual knowledge" (in Russian). *Materialy Simpoziuma po Logike Nauke* (Symposium Proceeding on Science of Logic). Kiev: Naukova Dumka.

Spirkin, A. G., Sazonov, B. V. 1964.
Methodological Problems of Studying Structures and Systems" (in Russian). *Voprosy Filosofii*, No. 1.

Trincher, K. S. 1965.
*Biologia i Informatsia* (Biology and Information), Moscow: Nauka Press.

Von Foerster, H. , 1960.
"On Self-Organizing Systems and Their Environments" In: Eds. Yovits, M. C. and Cameron, S. *Self-Organizing Systems*,

*References*

Proceedings of an Interdisciplinary Conference, May 3 and 6, 1959.
Vygotsky, L. S., 1934.
   *Myshlenie I Rech* (Thought and Speech). Moscow: OGIZ.

# Other Books by Vladimir A. Lefebvre

*What Are Consciousness, Animacy, Mental Activity and the Like?* Los Angeles: Leaf & Oaks Publishers, 2014.

*The Lectures on the Reflexive Game Theory*, Los Angeles: Leaf & Oaks Publishers, 2010.

*Research on Bipolarity and Reflexivity*, Lewiston: The Edwin Mellen Press, 2006.

*Algebra of Conscience*, 2nd enlarged edition (includes a new second part, "Moral Choice," not published before), Dordrecht: Kluwer, 2001.

*The Cosmic Subject*, Moscow: Russian Academy of Sciences Institute of Psychology Press, 1997.

Made in the USA
Columbia, SC
01 April 2021